INTRODUCING
ASTRONOMY

Other Titles in this Series:

Introducing The Planets and their Moons (2014)
Introducing Geology – A Guide to the World of Rocks (Second Edition 2010)
Introducing Geomorphology (2012)
Introducing Meteorology ~ A Guide to the Weather (2012)
Introducing Mineralogy (2014)
Introducing Oceanography (2012)
Introducing Palaeontology – A Guide to Ancient Life (2010)
Introducing Sedimentology (2014)
Introducing Tectonics, Rock Structures and Mountain Belts (2012)
Introducing Volcanology ~ A Guide to Hot Rocks (2011)

For further details of these and other Dunedin
Earth and Environmental Sciences titles see
www.dunedinacademicpress.co.uk

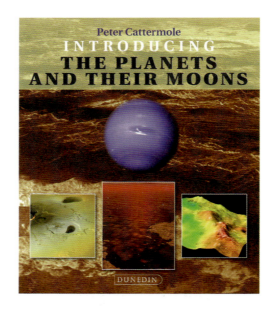

INTRODUCING
ASTRONOMY
A GUIDE TO THE UNIVERSE

Iain Nicolson

EDINBURGH ◆ LONDON

Published by Dunedin Academic Press Ltd

Head Office: Hudson House,
8 Albany Street, Edinburgh EH1 3QB

London Office: The Towers,
352 Cromwell Tower, Barbican, London EC2Y 8NB

www.dunedinacademicpress.co.uk

ISBNs
9781780460253 (Paperback)
9781780465159 (ePub)
9781780465166 (Kindle)

British Library Cataloguing in Publication data
A catalogue record for this book is available from the British Library

Design and pre-press production by Makar Publishing Production, Edinburgh
Printed and bound in Poland by Hussar Books

Contents

Acknowledgements

I am grateful to all of the many institutions and individuals who have kindly granted permission to reproduce images. I should also like to thank several reviewers for reading the manuscript and for their helpful and constructive comments and suggestions. Any errors or inadequacies that remain are, of course, my own responsibility.

My thanks are also due to those working with Dunedin Academic Press. In particular, I wish to thank Anthony Kinahan for initiating the project and for his unstinting support and advice; David McLeod, who has designed and typeset the book; copyeditor Anne Morton for her meticulous editing of the text and for her invaluable suggestions and advice; and the illustrator, Sandra Mather, for so skilfully translating my rough sketches into high-quality artworks and diagrams.

Finally, I thank my wife, Jean, for her support, encouragement and advice throughout the preparation of this book.

Iain Nicolson, June 2014

Note: I have tried to minimise the use of technical terms, but those that are necessary are explained in the Glossary and are highlighted in **bold** type on first appearance.

Cover Image credits:

Top, Saturn aurora: NASA, ESA, J. Clarke (Boston University), and Z. Levay (STScI);

Left hand, the Helix Nebula: NASA/JPL-Caltech;

Centre, a tower of cold gas and dust in the Eagle Nebula: NASA, ESA and the Hubble Heritage Team (STScI/AURA);

Right hand, Curiosity Rover on Mars : NASA/JPL-Caltech/Malin Space Science Systems;

Background, the Pinwheel galaxy: European Space Agency & NASA. Acknowledgements: Project Investigators for the original Hubble data: K.D. Kuntz (GSFC), F. Bresolin (University of Hawaii), J. Trauger (JPL), J. Mould (NOAO), and Y.-H. Chu (University of Illinois, Urbana); Image processing: Davide De Martin (ESA/Hubble); CFHT image: Canada-France-Hawaii Telescope/J.C. Cuillandre/Coelum; NOAO image: George Jacoby, Bruce Bohannan, Mark Hanna/NOAO/AURA/NSF.

List of tables and illustrations

Preface – the science of the universe

Astronomy is the science that investigates the universe around us. It is, arguably, the oldest of the sciences, for humankind must have been aware, long before written records began, of the Sun, Moon, stars and naked-eye planets and their movements in the sky. Gradually, sky watchers began to chart their positions, identify patterns in their behaviour, and use that information – on the one hand for practical purposes, to record the passage of time, the recurrence of the seasons, and as an aid to navigation – and on the other, with increasing sophistication over the centuries, to attempt to understand the nature of these objects and the scale of the universe in space and time.

For millennia, these studies were restricted to what could be seen directly by the unaided human eye. That changed, at the beginning of the seventeenth century, with the invention of the telescope, which enabled astronomers to see details on the surfaces of the Sun, Moon and planets, to detect objects far fainter than the eye alone could see and, therefore, to penetrate much deeper into space. Another key step, in the nineteenth century, was the invention of the spectroscope, a device that allowed astronomers to analyse in detail light arriving from distant objects, and to investigate their chemical and physical properties. In that same century, photography arrived on the scene, enabling astronomers to record permanent images and to detect objects that were too faint to be seen by the human eye even with telescopic aid. During the twentieth century and into the twenty-first, progressively larger telescopes, and advances such as electronic imaging, the computer analysis of data, and the ability to place telescopes in orbit beyond our atmosphere, and to send spacecraft to the planets and moons of our immediate cosmic neighbourhood, have utterly transformed the depth and breadth of the science.

But astronomy remains essentially an observational science. Whereas a chemist or physicist can set up an experiment in a laboratory, change the conditions and measure the outcome, astronomers cannot touch, or experiment directly with, stars or galaxies. Instead, they have to piece together their picture of the universe by detecting, measuring and analysing light and other forms of radiation that are coming our way from the depths of space. .

Part of astronomy's enduring fascination is that it grapples with fundamental questions concerning the origin, evolution and ultimate fate of planets, stars, galaxies and the universe as a whole, and touches on that most intriguing of conundrums: is life unique to the Earth or a widespread phenomenon? Another part of its great attraction is its accessibility: we can all go out and explore the beauty of the night sky for ourselves. This book outlines and explains what astronomy has to tell us about the nature of planets, stars, galaxies and the universe, and highlights some of the ways in which astronomers have arrived at this body of knowledge.

1 Our place in space

We look out into the vastness of the **universe** from the surface of planet Earth, a small rocky world that travels around the Sun (Fig. 1.1). The Sun is a typical **star** – a self-luminous globe of gas, composed mainly of hydrogen and helium, which generates its prodigious outpouring of energy by means of nuclear reactions that take place deep down in its interior. With a diameter more than a hundred times greater than that of the Earth, the Sun could comfortably contain well over a million bodies the size of our modest planet. It is overwhelmingly the dominant body in the **Solar System**, a set of bodies that consists of the Sun, eight **planets** and their natural **satellites** (moons), a host of smaller bodies, and

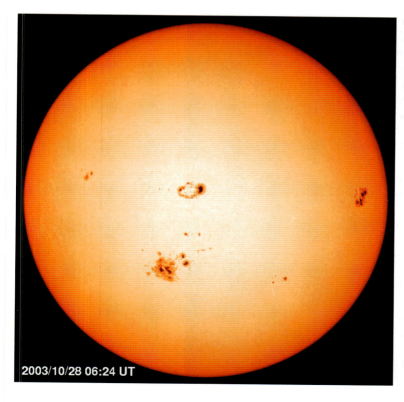

2003/10/28 06:24 UT

Figure 1.1 The Sun as imaged by the SOHO spacecraft on 28 October, 2003. The dark patches on its visible surface (the photosphere) are sunspots. Image: SOHO (ESA & NASA).

quantities of gas and dust, all of which revolve around the Sun, under the influence of its powerful gravitational pull.

Our nearest neighbour in space is the Moon, Earth's natural satellite. A barren, airless, rocky world, (Fig. 1.2) with about one quarter of the Earth's diameter, it revolves around our planet in a period of 27.3 days at a mean distance of 384,400 km, which is equivalent to about ten times the Earth's circumference. Whereas this is a distance that is relatively easy to visualize, it is often helpful to think about cosmic distances in terms of **light-travel time** – the time it would take a ray of light (the fastest-moving entity in the universe) to travel a particular distance. Moving at a speed of 300,000 kilometres per second, a ray of light takes 1.3 seconds to reach us from the Moon. By way of comparison, the Sun is about 400 times further away than the Moon, and a ray of light takes 8.3 minutes to cross the 149,600,000 kilometre gulf that separates us from this, our neighbourhood star.

In order of distance from the Sun, the planets are: Mercury, Venus, the Earth, Mars, Jupiter, Saturn, Uranus and Neptune. The

Figure 1.2 Earth's nearest neighbour, the Moon, showing its dark plains and lighter, cratered highlands. Image credit: C.R. Lynds, KPNO/NOAO/NSF.

innermost planet, Mercury, orbits at about two-fifths of the Earth's distance and takes just 88 days to travel round the Sun, whereas the most distant planet – Neptune – is about 30 times further from the Sun than we are, and takes nearly 165 years to crawl round its much larger orbit. The light-travel time from the Sun to Neptune is just over four hours.

The nearest star, apart from the Sun itself, is a dim red one called Proxima Centauri (a faint companion to the bright naked eye star, Alpha Centauri (Fig. 1.3)), which is far less powerful than our Sun and much too faint to be seen with the unaided eye. Lying at a distance of more than 40 trillion kilometres (40,000,000,000,000 km, or $4×10^{13}$ km) (Table 1.1), it is more than a quarter of a million times further away from us than is the Sun, and its light takes 4.2 years to traverse that great distance. The distance travelled by light in one year – about 9.46 trillion kilometres – is known as a **light-year;** expressed in those terms, Proxima Centauri lies at a distance of 4.2 light years.

Figure 1.3 Part of the Milky Way. Of the two bright stars to the left of centre, the one further to the left is Alpha Centauri, the nearest naked eye star. Image credit: ESO.

Table 1.1 Powers of ten. A convenient way of expressing very large (or very small) numbers is to use index notation, or 'powers of ten', where 10^n (ten to the power n) denotes the number 1 followed by n zeros, and 10^{-n} (ten to
the power minus n) represents 1 divided by (1 followed by n zeros). For example, using this notation, the number
9,460,000,000,000 (9.46 trillion) would be written as 9.46×10^{12}, whereas 0.0000000000055 (5.5 trillionths) would be
written as 5.5×10^{-12}.

Number	Decimal form	Index notation
one hundred	100	10^2
one thousand	1,000	10^3
one million	1,000,000	10^6
one billion	1,000,000,000	10^9
one trillion	1,000,000,000,000	10^{12}
one hundredth	0.01	10^{-2}
one thousandth	0.001	10^{-3}
one millionth	0.000001	10^{-6}
one billionth	0.000000001	10^{-9}
one trillionth	0.000000000001	10^{-12}

A scale model may help to give a feel for these distances. If we were to represent the Sun by a large grapefruit 14 cm in diameter, the Earth would be a small pinhead, 1.25 mm in diameter, located at a distance of 15 metres from the model Sun. Jupiter – the largest of the planets – would be a 1.4 cm (small) grape at a distance of 78 metres and Neptune, a 5 mm berry at a distance of about 450 metres. On this scale, the nearest star (Proxima Centauri) would be represented by a small orange at a distance of about 4000 km. If the model Sun were located in London, the nearest star would be in Baghdad, Iraq; if it were in New York, the nearest star would be in Los Angeles.

The stars are all self-luminous gaseous bodies, similar in general nature to the Sun. They appear so much fainter only because they are so very much further away. Although broadly similar in nature to our Sun, stars exhibit a very wide range of properties. At one end of the scale are stars which are hundreds of times larger and tens of thousands, or even millions, of times more luminous than the Sun,

while at the other end of the range are dwarf stars, with less than a hundred thousandth of the Sun's light output, some of which (white dwarfs) are comparable in size with planet Earth. Some are more massive than the Sun, some less massive; some are hotter, some are cooler, some younger, and some older. In recent years, observations have revealed that many stars have planets (**exoplanets**) revolving around them. Our Solar System is certainly not unique.

All of the stars that we can see with the unaided eye on a clear, dark night are members of our **galaxy**, a huge star system which contains, in total, at least 200 billion stars, spread out in a flattened disc-shaped aggregation that measures more than 100,000 light-years in diameter. The Solar System lies in the galactic suburbs, some 27,000 light-years away from the galactic centre (the centre of the galaxy).

Our galaxy (commonly known as the **Milky Way** galaxy) is one of well over a hundred billion galaxies of various shapes, sizes and

Figure 1.4 The Andromeda galaxy (M31), which is similar to, but larger than, our own galaxy, lies at a distance of 2,500,000 light-years. Image credit: Bill Schoening, Vanessa Harvey/REU program/NOAO/AURA/NSF.

properties that lie within range of present-day telescopes. The nearest large one – the **Andromeda galaxy** (so-called because, viewed from the Earth, it lies within the constellation of Andromeda) – is 2.5 million light-years away (Fig. 1.4). Otherwise known as M31 (object number 31 in the catalogue of fuzzy nebulous objects published in 1781 by French astronomer Charles Messier), this galaxy is sufficiently bright – despite its great distance

– to be glimpsed with the naked eye as a faint misty patch of light. For most people, it is by far the most remote object that can be seen without optical aid.

The most distant galaxies are so far away that their light has taken more than 13 billion years to reach us, a period of time that is comparable to the age of the universe, and about three times as great as the age of planet Earth (Fig. 1.5). When we detect the faint light arriving today from these most remote of objects, we are seeing them as they were,

Figure 1.5 The image shows the Hubble Ultra Deep Field 2012, which contains some of the most distant galaxies ever seen, some of which are so remote that their light has taken 13 billion years to reach us. Image credit: NASA, ESA, R. Ellis (Caltech), and the HUDF 2012 Team.

billions of years ago, when that light set out on its journey through space. Because light travels at a finite speed, our view of distant objects is always 'out of date'. We see the Moon as it was 1.3 seconds ago, and the Sun as it was 8.3 minutes ago. If the Sun were to be annihilated at the instant at which you are reading these words, you would not become aware of the fact until 8.3 minutes from now. We see Neptune as it was some 4 hours ago, Proxima Centauri as it was 4.2 years ago, the Andromeda galaxy as it was 2.5 million years ago, and the most remote galaxies as they were about 13 billion years ago. Although the finite speed of light implies that we cannot see every object in the universe as it is 'now', the compensating benefit is that, by probing further and further into the remote recesses of the cosmos, we can see what the universe was like in the distant past, and examine how it has changed and evolved over the past 13–14 billion years. The scale of the **observable universe** (that part of the universe which, in principle, we can see from planet Earth) in space and time is truly immense.

2 The changing sky

Seen from a good dark site, when the sky is clear and the Moon is not around, the night sky presents a glorious panorama of stars, the brightest glowing like sparkling jewels set against a velvet black background. Under these conditions, it is easy to appreciate why the ancient sky-watchers of thousands of years ago came to regard the sky as a hemispherical dome set above what, in those times, was believed to be a flat Earth.

The early watchers of the sky identified patterns among the stars – some more obvious than others – that were constant and unchanging; the same stars remained always in the same positions relative to their neighbours. These patterns, or **constellations**, were often named after creatures or mythological characters. Each of the early civilizations identified their own distinctive star patterns, but the majority of the constellations that are internationally recognized today are based on those which were identified and named by the ancient Greeks, though their formal names are quoted in Latin form. Conspicuous examples are Ursa Major (the Great Bear – which is commonly known as 'the Plough' or 'the Big Dipper' (Fig. 2.1), Orion (the Hunter), Cygnus (the Swan), and Leo (the Lion)). The constellations of the far southern skies, which could not be seen by astronomers in European latitudes, were identified and named later. Notable southern constellations include Centaurus (the Centaur) and Crux Australis (the Southern Cross). The entire sky has been divided into 88 constellations, with boundaries that were formally defined in 1930 by the International Astronomical Union (the organization which, among many other things, establishes the agreed names for astronomical bodies).

By convention (although there are a few anomalous cases), the brightest star in a constellation is designated by the Greek letter, Alpha (α), the second brightest, Beta (β), the third brightest, Gamma (γ), and so on. For example, the brightest star in Leo is designated as α Leonis. The brighter stars also have individual proper names, many of which were assigned by the highly accomplished Arab astronomers. Notable examples are Aldebaran (α Tauri), the brightest star in the constellation of Taurus (the Bull) and Deneb (α Cygni), the brightest star in Cygnus. The brightest star in the entire sky (apart from the Sun itself) is Sirius (α Canis Majoris).

Constellations have no physical significance. The individual stars that make up a constellation may lie at very different distances from us; for example, some of the stars which make up the splendid constellation Orion are further from each other than we are from some of them. They appear relatively close together in the sky, and make up their particular pattern, simply because they lie in similar directions as viewed from the Earth.

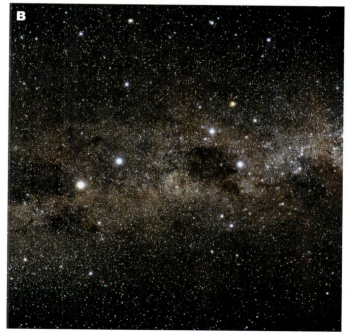

Figure 2.1 A, Ursa Major: the seven brightest stars make up the familiar pattern known as the Plough, or Big Dipper. Image credit: A. Fujii. **B**, Centaurus and Crux Australis (the Southern Cross): the two bright stars in Centaurus, Alpha and Beta Centauri (towards the left of the image) point towards the kite-shaped pattern of the Southern Cross (right of centre). Image credit: ESO/S. Brunier.

The rotating sky

The Earth rotates around its axis from west to east. If you were hovering above the North Pole, you would see the Earth spin beneath you in an anticlockwise direction, whereas from above the South Pole, the Earth would be seen to turn in a clockwise direction. To an observer on the surface of the rotating Earth, it seems as if the Sun, Moon, stars and planets are revolving round our world from east to west. Indeed, the ancient Greek astronomers of two millennia ago believed that the stars were indeed fixed to a huge sphere that rotated round the Earth once a day. Although we now know that the stars are remote suns lying at vast, and very different, distances, we still find it convenient, when describing their positions and motions, to maintain the fiction that they are attached to a sphere – the **celestial sphere** – that rotates around our planet (Fig. 2.2).

The Earth's axis of rotation, extended into space, meets this imaginary sphere at the north and south **celestial poles**, and the Earth's equator, projected into space, bisects the celestial sphere along a circle that is called the **celestial equator**. Viewed from the Earth's north (or south) pole, the north (south) celestial pole would be directly overhead. If you were located on the Earth's equator, the celestial equator would pass directly overhead. In the northern hemisphere, there is one reasonably bright star – Polaris – which lies within one degree of the true north celestial pole and is, for that reason, often referred to as the Pole Star or the North Star. The southern hemisphere does not possess a similarly bright star anywhere close to its celestial pole.

At any instant, an observer on the Earth's surface can see half of the celestial sphere, the other half being hidden below the horizon. As the Earth turns round, the celestial sphere

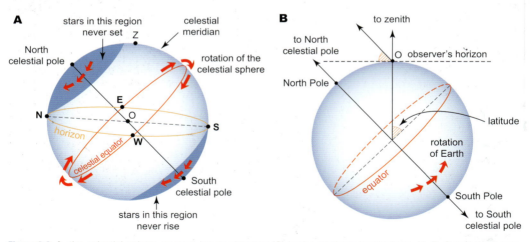

Figure 2.2 A, the celestial sphere as seen by an observer (O) in the northern hemisphere. N, E, S and W denote, respectively, the north, east, south and west points of the horizon, and Z, the zenith. **B**, the observer's location on the Earth's surface. The angle between the horizon and the north celestial pole (the altitude of the celestial pole) is equal to the observer's latitude.

appears to rotate from east to west and stars move across the sky, parallel to the celestial equator, tracing out circles around the celestial pole. For an observer located at the north (or south) pole, the celestial equator coincides with the horizon and stars move round parallel to the horizon; half of the celestial sphere is always above the horizon and the other half is permanently hidden. In contrast, when viewed from the equator, the celestial equator crosses the horizon perpendicularly at the east and west points and passes through the **zenith** (the point that is directly overhead), whereas the celestial poles coincide with the north and south points of the horizon. Although an observer at the equator can only see half of the celestial sphere at any particular instant, the rotation of the Earth ensures that each part of the sphere can be seen at one time or another, and that all of the stars rise and set.

For an observer located somewhere between the equator and a pole, some stars – called **circumpolar stars** – remain above the horizon at all times, tracing out circles centred on the celestial pole but never dipping below the horizon (Fig. 2.3). Other stars rise and set, while some remain permanently below the horizon. Which stars particular observers can see depends on their latitude on the Earth's surface and on how far north or south of the celestial equator the individual stars lie.

Figure 2.3 Star trails: this image, taken from La Silla Observatory, Chile, shows the apparent rotation of stars around the south celestial pole. During a long exposure, the apparent rotation of the celestial sphere draws out their images into curved trails. Image credit: Iztok Boncina/ESO.

The apparent annual motion of the Sun

The stars that we see in the night sky are on the opposite side of the celestial sphere from the Sun, the others being hidden by the brightness of the daytime sky. As the Earth moves round the Sun, its night hemisphere faces in a steadily changing direction so that different stars and constellations become visible at different times of the year. For example, the northern hemisphere sky in winter is dominated by the magnificent constellation Orion, which acts as a prominent signpost from which to find other neighbouring constellations, whereas prominent summer constellations include Cygnus, Lyra and Aquila, the brightest member stars of which (Deneb, Vega and Altair, respectively) make up a conspicuous pattern, commonly known as 'the Summer Triangle' (Fig. 2.4). Each constellation rises (and sets) about four minutes earlier on each consecutive night – about two hours earlier on each successive month – returning, after a full year, to its original position in the night sky.

Figure 2.4 Constellations: **A**, Orion: the three stars near the centre form Orion's belt; the bright red star towards the top is Betelgeuse, and the bright star near the bottom is Rigel. **B**, the bright stars Deneb (top of image), Vega (above and right of centre), and Altair (below, left of centre), in the constellations Cygnus, Lyra and Aquila, respectively, make up a conspicuous pattern, known as the Summer Triangle. Image credit: A, Nik Szymanek; B, A. Fujii.

If we could see the stars in daytime, we would see the Sun against a starry background. In one month, the Earth travels about one-twelfth of the way round the Sun and moves through an angle of about 30 degrees (one-twelfth of 360°) as viewed from the Sun. Viewed from the Earth, the Sun appears to move through an angle of 30° relative to the background stars. In the course of a year, the Earth makes one complete circuit of the Sun, and the Sun, as seen from the Earth, makes one complete circuit of the celestial sphere, tracing out a path that is called the **ecliptic** (Fig. 2.5).

If the Earth's axis were perpendicular to the plane of its orbit, the celestial equator (which lies in the plane of the Earth's equator) and the ecliptic (which lies in the plane of the Earth's orbit), would coincide. But because the Earth's axis is tilted from the perpendicular by an angle of about 23.5 degrees, and the Earth's equator is tilted to the plane of its orbit by this same angle, the ecliptic intersects the celestial equator at an angle of 23.5 degrees, crossing the celestial equator at two points: the **vernal equinox**, which is the point at which the Sun crosses the celestial equator from south to north, and the **autumnal equinox** – the point at which the Sun crosses the celestial equator from north to south.

Celestial coordinates

A circle that passes through a particular place and through the north and south poles, and which crosses the equator at right angles, is called a **meridian**. The location of a place on the Earth's surface is specified by its latitude and longitude. **Latitude** is the angular distance north (N) or south (S) of the equator, measured along a meridian (equivalently, it is the angle between the equator and the place as viewed from the centre of the Earth). It can take any value from 0° (at the equator) to 90° (at a pole). **Longitude** is the angle between the Greenwich meridian (the meridian that passes through the old Royal Observatory at Greenwich, England) and the meridian that passes through the place of interest. Longitude

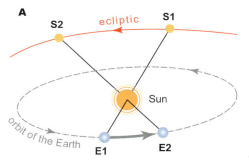

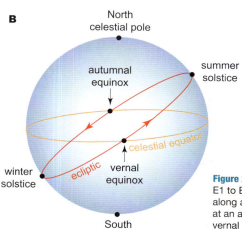

Figure 2.5 The ecliptic: **A**, as the Earth moves along its orbit from E1 to E2 the observed position of the Sun shifts from S1 to S2 along a path that is called the ecliptic. **B**, the ecliptic is inclined at an angle to the celestial equator, crossing it at two points – the vernal equinox and the autumnal equinox.

takes values between 0° and 180° measured east (E) or west (W) from the Greenwich meridian, with 180° E coinciding with 180° W.

In an analogous fashion, a circle passing through both celestial poles and a particular star, or some specific point on the celestial sphere, is called an **hour circle** – the hour circle that passes through an observer's zenith, and which intersects the horizon at right angles, at its north and south points, being called a **celestial meridian**. The position of a star is specified by its **right ascension** and **declination**. Declination (dec) is the angle measured north (+) or south (–) between the celestial equator and the star. It takes values from 0° (for a star on the celestial equator) to +90° (for a star at the north celestial pole) or to –90° (for a star located at the south celestial pole). Right ascension (RA) is the angle measured eastwards – anticlockwise in the northern hemisphere – from the hour circle passing through the vernal equinox to the hour circle passing through the star. RA takes values between 0° and 360° but, by convention, is normally expressed in time units (hours, minutes and seconds), from 0 to 24 hours. This is because the Earth, and hence the sky, rotates through an angle of 360° in 24 hours, and therefore turns through 15° in 1 hour. One hour of RA is equivalent to 15°, 6 hours to 90°, and so on (Fig. 2.6).

The position of a star at a particular instant can also be described by its **altitude** and **azimuth**. Altitude is the angle between the horizon and the star, measured perpendicular to the horizon. Azimuth is the angle, measured parallel to the horizon, between the star and the point on the horizon that is vertically below the star; it is measured in a clockwise direction from the north point of the horizon,

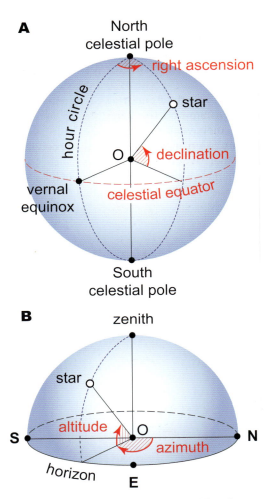

Figure 2.6 Celestial coordinates: A, the position of a star on the celestial sphere is specified by its right ascension and its declination. B, its position as seen by observer O at a particular instant can be described by two angles – altitude and azimuth.

taking values between 0° and 360°. A star that is due east would have an azimuth of 90°, due south 180°, due west 270°, and so on. The altitude and the azimuth of a star both change continuously (except at the north and south poles, where stars move parallel to the horizon and their altitudes, therefore, do not change). Apart from circumpolar stars, which are always above the horizon, all stars rise on the eastern side of the meridian (somewhere between azimuth 0° and 180°), reach their greatest altitudes (**upper transit**, or **culmination**) when they cross the meridian, and then eventually set on the western side of the meridian somewhere between azimuth 180° and 360°.

The altitude of the celestial pole is always equal to the observer's latitude; for example, viewed from the north pole (latitude 90° N) the north celestial pole is vertically overhead (altitude 90°), from the equator it is down on the horizon (altitude 0°), and from Edinburgh (latitude 56° N), its altitude is 56°. If the angular distance between the north (south) celestial pole and a particular star is precisely equal to the observer's latitude, that star will trace out a circle which, at its lowest point, *just* touches the north (south) point of the horizon. Any star that lies within this circle will be circumpolar, and will never set. For example, the seven principal stars of Ursa Major (the Plough or Big Dipper) are circumpolar at all latitudes further north than 41° N, whereas Crux Australis (the Southern Cross) is circumpolar for all latitudes further south than about 35° S.

Seasonal changes

Because the Earth's axis is tilted and points in a fixed direction in space (apart from a very slow periodic change called **precession**, which takes place over a period of 25,800 years), when the Earth is one side of the Sun (in December), its south pole is continuously illuminated, and its north pole is in shadow; six months later (in June), the north pole is illuminated and the south pole is in darkness.

When the Sun is at the vernal equinox, on or around 21 March each year, it is on the celestial equator, and its declination is 0°. It therefore rises due east and sets due west, passing vertically overhead at noon for an observer on the Earth's equator; viewed from either pole, the Sun will be sitting on the horizon. At every point on the Earth's surface (apart from at the poles themselves) the Sun will be above the horizon for 12 hours and below the horizon for 12 hours so that day and night will be of equal duration. Three months later – on or around 21 June – the Earth will have moved one quarter of the way round its orbit and the Sun will then be 23.5 degrees north of the celestial equator, so that everywhere within 23.5 degrees of the north pole (i.e., within the Arctic Circle) will experience continuous daylight as the planet rotates. Conversely, every place within the Antarctic Circle (within 23.5 degrees of the south pole) experiences continuous night. On this date, which (with northern hemisphere chauvinism) is known as the **summer solstice**, everywhere in the northern hemisphere has more than 12 hours of daylight, and less than 12 hours of darkness, in each 24-hour period, whereas everywhere in the southern hemisphere has less than 12 hours of daylight. The higher the latitude, the more pronounced the effect (Fig. 2.7).

After a further three months, on or around 22 September, the Sun crosses the equator from north to south, and day and night are again of equal duration all over the planet. The Sun then heads south of the celestial equator until, at the **winter solstice** (around 22 December) it is as far south as it can get

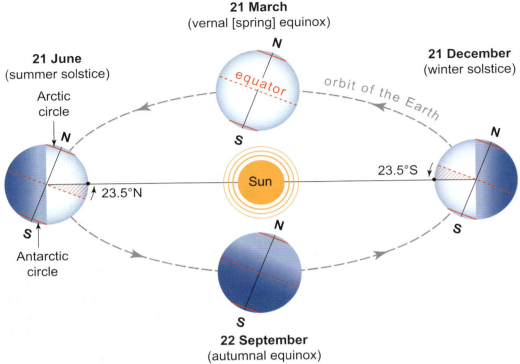

21 March
(vernal [spring] equinox)

21 June
(summer solstice)

Arctic
circle

23.5°N

Antarctic
circle

21 December
(winter solstice)

orbit of the Earth

23.5°S

Sun

22 September
(autumnal equinox)

Figure 2.7 The seasons: the Sun is vertically overhead at noon at latitude 23.5° N on 21 June (the summer solstice), at 23.5°S on 21 December (the winter solstice), and at the equator on the two equinoxes (March and September). The Arctic is continuously illuminated at the summer solstice, whereas the Antarctic is continuously illuminated at the winter solstice.

(23.5° S). On this date, everywhere within the Antarctic Circle is bathed in continuous sunlight and everywhere within the Arctic Circle is in darkness. At this time, every place in the southern hemisphere experiences more than, and everywhere in the northern hemisphere less than, 12 hours of daylight.

The noon altitude of the Sun is affected by its changing declination, which ranges from +23.5° on midsummer's day to –23.5° on midwinter's day. From Edinburgh, for example, its noon altitude ranges between 57.5° in mid-summer to just 10.5° in mid-winter. In the northern hemisphere, the Sun rises due east (and sets due west) at the equinoxes, and rises progressively further north of east (and sets further west of south) between the dates of the vernal equinox and the summer solstice. After the summer solstice, the rising point begins to migrate back towards due east (and the setting point towards due west), where it arrives on the autumnal equinox. Between the autumnal equinox and the winter solstice, the rising point moves progressively further south of east (and the setting point further south of west). Thereafter, the rising and setting points begin to migrate back from their southern extremes. The term 'solstice', which means 'standstill of the Sun' reflects the fact that the daily shift in the Sun's rising and setting points comes briefly to a halt, then reverses, on those occasions.

Telling the time

The rotation of the Earth determines the duration of the day, but there are several variants of the term, 'day' – the sidereal day, the apparent solar day and the mean solar day. The **sidereal day** is the time interval between two successive upper transits of the vernal equinox, or of a particular star, and is precisely equal to the rotation period of the Earth – the time taken for the Earth to turn through an angle of 360° relative to the background of distant stars. It is divided into 24 hours of **sidereal time**. The Earth turns through an angle of 15 degrees in each sidereal hour, 15 arcminutes (a quarter of a degree) in each sidereal minute and 15 arcseconds in each sidereal second. The **apparent solar day** is the time interval between two successive noons – two successive upper transits of the Sun – and is divided into 24 hours of **apparent solar time**.

Two factors conspire to cause the duration of successive solar days to vary slightly in a periodic way – the tilt of the ecliptic and the elliptical nature of the Earth's orbit. Because of the tilt of the ecliptic relative to the celestial equator, the daily motion of the Sun relative to the background stars consists of two parts: north–south motion perpendicular to the celestial equator and west–east motion (change in right ascension), parallel to the celestial equator. Around the time of the equinoxes the rate of change of declination is greatest, and the rate of change of right ascension is least; in contrast the rate of change of declination is least, and the rate of change of right ascension greatest, around the time of the solstices. These variations cause small, cumulative, changes in the time at which the Sun reaches upper transit (noon). A further variation in the apparent day length is caused by the fact that the Earth moves more quickly (and the apparent motion of the Sun is therefore greater) along its orbit when it is closer to the Sun and more slowly when it is further away.

The mean duration of the solar day, averaged over the whole year, defines the mean solar day, which is divided into 24 hours of **mean solar time** (**mean time**). Formerly (and still commonly) known as Greenwich Mean Time (GMT), **Universal Time** (**UT**) is a system of time measurement based on the mean solar time as measured by an observer located on the Greenwich meridian (the meridian that passes through the old Royal Observatory in Greenwich, England; this is the basis of everyday civil time.

The mean solar and sidereal days are not identical because, in addition to rotating around its axis, the Earth is revolving round the Sun, moving along its orbit at a rate (as viewed from the Sun) of just under one degree per day. A given star will return to upper transit on an observer's meridian after one sidereal day, during which time the Earth has rotated through an angle of 360°. But, because the Earth, during this interval of time, will have moved about one degree along its orbit, the Sun will appear to have travelled about one degree eastward along the ecliptic. Consequently, the Earth has to rotate through an additional degree before the Sun is again aligned with the observer's meridian and, because it takes the Earth about four minutes to turn through that extra degree, the solar day is about four minutes longer than the sidereal day (Fig. 2.8). Expressed in mean solar time, the axial rotation period of the Earth, and the duration of the sidereal day, is 23h 56m. Over a complete year the Sun rises and sets 365 times, but a given star will rise and set 366 times. Consequently, each individual star crosses the observer's meridian about four minutes earlier on each consecutive night until, after one complete year, it again crosses the meridian at the same instant of mean time.

Because of the cumulative effect of minor variations in the axial rotation of the Earth, time today is determined not by the Earth's rotation, but by comparing the time recorded on a set of ultra-precise atomic clocks based in a number of laboratories around the Earth; this time, which is the precise basis of scientific time measurement and civil time, is called **coordinated universal time** (**UTC**).

Grappling with the calendar
The month derives originally from the time interval between two successive New Moons (*see* Chapter 3), and the year is, in essence, the orbital period of the Earth. More specifically, the time interval between two successive occasions on which the Sun arrives at the

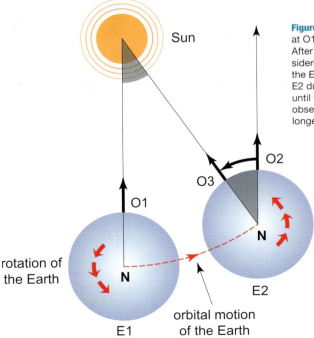

Figure 2.8 Solar and sidereal days: An observer at O1 is looking directly towards the Sun at noon. After one complete rotation of the Earth (one sidereal day), the observer is at O2. Because the Earth has moved along its orbit from E1 to E2 during this time interval, noon will not recur until the Earth has rotated further, bringing the observer to O3. Consequently, the solar day is longer than the sidereal day.

Sun

O2

O3

O1

N

N

rotation of the Earth

E2

E1

orbital motion of the Earth

vernal equinox, which defines the recurrence of the seasons and is called the **tropical year**, is 365.2422 days, approximately one quarter of a day more than 365 days. Constructing a workable calendar has been a challenge to humankind for millennia, because the year is not a whole number of days, and the time interval between two successive New Moons (29.53 days) is neither a whole number of days nor a simple fraction of a year.

The Julian calendar, which incorporates amendments to the Roman calendar which were made in 45 BC by Julius Caesar, took account of the quarter day by adding an extra day to every fourth year, so introducing the idea of the 'leap year'. According to the Julian calendar, a year would be a leap year (and an extra day would be added to February) when the last two digits of the year number were divisible, without remainder, by 4; for example, 2012 was a leap year (12 ÷ 4 = 3, a whole number), but 2009, 2010, and 2011 were not. Subsequently, in a pragmatic attempt to keep days, months and years as closely in step as possible, the year was divided into seven months of 31 days, four of 30 days and one – February – of 28 days (except in leap years, when it contains 29); this gave a cycle of three consecutive 365 day years followed by one of 366 days, which made the mean length of the calendar year equal to 365.25 days. However, even this degree of complexity is insufficient – over very long periods of time – to keep the calendar year and the seasons in precise step. The Gregorian calendar that we use today incorporates reforms to the Julian calendar that were introduced in 1582 by Pope Gregory XIII. Gregory decreed that century years – ending in 00 – would be leap years only when the year is divisible by 400 without remainder; thus the year 2000 was a leap year (2000 ÷ 400 = 5, a whole number) whereas 1700, 1800, 1900 were not. With Gregory's amendment in place, the average length of the calendar year differs from the tropical year by just one day in 3000 years.

If intelligent beings exist on other planets elsewhere in the galaxy, who knows what problems they will have had to grapple with in order to construct their calendars?

3   Planets and orbits

Since the dawn of recorded history human-kind has been aware of the presence of five star-like objects which, although they rise in the east and set in the west just like the stars, slowly – night by night and month by month – change their positions relative to the starry background. These 'wandering stars' came to be known as **planets** (the word 'planet' derives from the Greek word 'planetes', meaning 'wanderer'). Two of these planets – Mercury and Venus – are always relatively close to the Sun in the sky and can only be seen for up to a few hours either in the morning sky, before sunrise, or in the evening sky after sunset. The other three – Mars, Jupiter and Saturn – when best placed, can remain visible all night long. For most of the time, they move slowly in a west to east direction (**direct motion**) relative to the background stars, but occasionally they will stop, run back for a while in the opposite direction (**retrograde motion**), and then revert to direct motion once again. On these occasions, over a period of a few weeks, they trace out elongated loops in the sky relative to the starry background.

We now know that this behaviour is caused by the relative motion of the Earth and the other planets as they move along their **orbits** round the Sun, but it took millennia for this realization to come. To the ancient Greeks, in the time of Aristotle (4th century BC), it seemed self-evident that the Earth was located at the centre of the cosmos, and that everything else – Sun, Moon, planets and stars – revolved around our world. Brought to its peak of development in the second century AD by Claudius Ptolemaeus (Ptolemy), the Earth-centred **geocentric system** went largely unchallenged until the early part of the six-teenth century when Polish cleric Nicolaus Copernicus asserted that the motions of celestial bodies could better be explained by assuming that the Earth rotates on its axis and, together with all the other planets, revolves around the Sun. Copernicus published his **heliocentric** (Sun-centred) theory in 1543, but it still failed to provide a complete explanation of planetary motion because, like the ancient Greeks, he had assumed that the orbits of the Earth and planets were circular.

It was not until the beginning of the sev-enteenth century that German astronomer Johannes Kepler – after painstaking detailed analysis of records of planetary positions gathered over many years by the Danish observer Tycho Brahe – made the crucial dis-covery that each planet moves round the Sun, not in a circle, but along an *elliptical* path. **Kepler's laws of planetary motion**, which were published between 1609 and 1618, are as follows:

First law: Each planet travels round the Sun in an elliptical orbit, with the Sun at one focus of the ellipse.

An **ellipse** is an oval figure. Its maximum diameter is called the **major axis** and its minimum diameter, the minor axis. There are two points, or foci, which are located on the major axis on either side of the centre of the ellipse, and the sum of the distances from the two foci to any point on the ellipse is constant. The greater the separation of the foci, the more elongated the ellipse. Because the Sun is located at one focus, the planet's distance from the Sun varies as it moves around its orbit. The point of closest approach is called **perihelion**, and the point at which it is furthest from the Sun is called **aphelion** (Fig. 3.1).

Second law: The **radius vector** (the line joining the Sun to the planet) sweeps out equal areas in equal times.

This implies that the speed of a planet varies as it moves along its orbit; it goes faster when closer to the Sun and more slowly when further away (Fig. 3.1B).

Third law: The square of a planet's **orbital period** is directly proportional to the cube of its mean distance from the Sun.

The term 'mean distance' in this context refers to the **semi-major axis** of the planet's

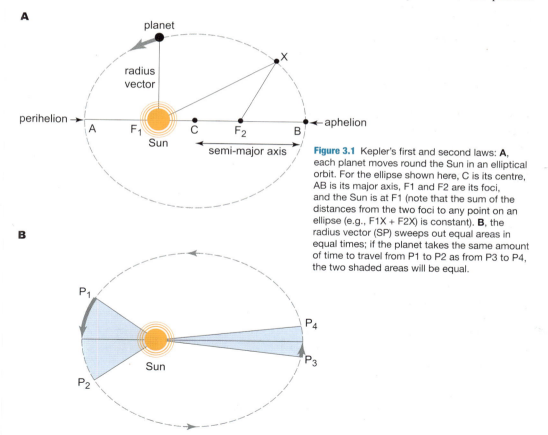

A

B

Figure 3.1 Kepler's first and second laws: **A,** each planet moves round the Sun in an elliptical orbit. For the ellipse shown here, C is its centre, AB is its major axis, F1 and F2 are its foci, and the Sun is at F1 (note that the sum of the distances from the two foci to any point on an ellipse (e.g., F1X + F2X) is constant). **B,** the radius vector (SP) sweeps out equal areas in equal times; if the planet takes the same amount of time to travel from P1 to P2 as from P3 to P4, the two shaded areas will be equal.

orbit – half of the length of the major axis. The semi-major axis of the Earth's orbit, which has a value of 149,600,000 km, is used as a unit of measurement, called the **astronomical unit** (AU). Expressed in these terms, the semi-major axis of the Earth's orbit is 1.00 AU. If distance is expressed in AU and time in years, Kepler's third law can be written down as 'period squared equals mean distance cubed'. For example, a planet at a mean distance of 4 AU would have an orbital period, in years, equal to the square root of 4^3 ($4 \times 4 \times 4 = 64$, and the square root of 64 is 8; therefore the planet's period would be 8 years).

Planetary movements, configurations and alignments

Planets that are closer to the Sun than is the Earth (i.e. Mercury and Venus) are called **inferior planets**, while those that are further away (Mars, Jupiter, Saturn, Uranus and Neptune) are known as **superior planets** (Fig. 3.2).

An inferior planet overtakes the Earth at regular intervals, and is said to be at **inferior conjunction** (the term **conjunction** means a close alignment between two celestial bodies) when it passes between the Sun and the Earth. After passing through inferior conjunction,

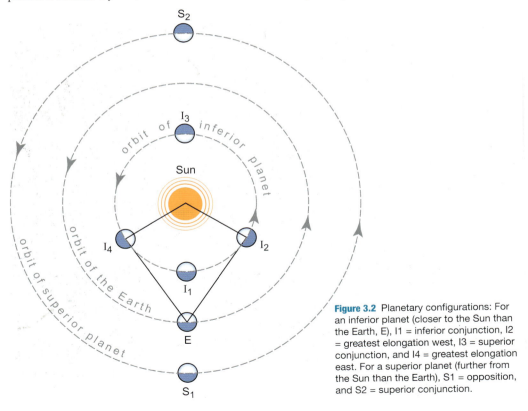

Figure 3.2 Planetary configurations: For an inferior planet (closer to the Sun than the Earth, E), I1 = inferior conjunction, I2 = greatest elongation west, I3 = superior conjunction, and I4 = greatest elongation east. For a superior planet (further from the Sun than the Earth), S1 = opposition, and S2 = superior conjunction.

the planet moves to the west of the Sun and becomes visible low in the eastern sky just before sunrise. Thereafter, the angle between the Sun and the planet (the **elongation**) increases until it reaches a maximum (greatest elongation west) at which point the line of sight from the Earth to the planet looks along a tangent to the planet's orbit. Greatest elongation for Venus is about 47°, whereas for Mercury (which has a markedly elongated elliptical orbit) it can take values ranging from 17° to 28°. The planet's elongation then begins to decrease, and continues so to do until it eventually passes behind the Sun, at

which point it is at **superior conjunction**. Thereafter, the planet emerges on the eastern side of the Sun and starts to become visible in the western sky after sunset. Its elongation then increases to a maximum (greatest elongation east) before starting to decrease as the planet begins to catch up once again with the Earth, heading once more towards inferior conjunction.

If an inferior planet's orbit lay exactly in the plane of the Earth's orbit, it would pass directly in front of the Sun each time it arrived at inferior conjunction, and on these occasions would be seen as a small dark disc

Figure 3.3 A multiple planetary conjunction, imaged on 1 May 2011. The Moon is on the lower left. Venus is the highest and brightest planet, with Mercury below and to its right. To the right of the Moon, and slightly lower, is Mars, with Jupiter slightly higher to its right. Image credit: G. Hudepohl (atacamaphoto.com)/ESO.

Figure 3.4 Transit of Venus: here, the dark disc of the planet is crossing the edge of the solar disc at the end of the transit of 6 June 2012. Image credit: NASA/SDO.

advancing across the face of the Sun. Such an event is called a **transit** (Fig. 3.4). In practice, because the orbits of Mercury and Venus are both inclined at an angle to the Earth's orbit, each will usually pass a little way above, or a little way below, the Sun, and a transit will occur only if inferior conjunction occurs when the planet is at, or very close to, one of the **nodes** of its orbit (one of the points at which it crosses the Earth's orbital plane). Transits of Mercury are not particularly rare; the last two took place in 2003 and 2006, and

the next two will occur on 9 May 2016 and 11 November 2019. Transits of Venus are much less frequent: the last two occurred in 2004 and 2012, but the next two will not take place until more than a century hence – on 11 December 2117 and 8 December 2125.

The Earth overtakes a superior planet at regular intervals. At that instant the Sun, the Earth and the planet come into line and, seen from the Earth, the planet is on the opposite side of the sky from the Sun. The planet is then said to be at **opposition**. When at opposition a superior planet is at its closest to us and culminates at midnight. The Earth then moves ahead of the planet and the planet lags further behind, drifting closer to the Sun in the sky and setting progressively earlier, until it passes behind the Sun and reaches superior conjunction. Thereafter, as the Earth begins once again slowly to catch up with it, the planet emerges on the west side of the Sun and becomes visible in the eastern sky just before sunrise. As its elongation increases, it rises progressively earlier, eventually rising before midnight and then moving inexorably towards its next opposition.

Because all of the planets travel round the Sun in the same direction, the usual motion of a planet relative to the background stars is from west to east but, for a few weeks round about the time of opposition, when the Earth is catching up with, then moving ahead of, a superior planet, the planet will, for a time, seem to be moving 'backwards' in a retrograde fashion before resuming its normal, direct, motion. The planet does not really move 'backwards'; it simply appears to do so because we are overtaking it.

Sidereal and synodic periods

The time that a planet takes to make one complete circuit around the Sun relative to the distant background stars is called its **sidereal orbital period** (or **sidereal period**). The sidereal periods for the planets range from 88 days, for Mercury, to 164.8 years for Neptune. The time interval between two successive similar configurations (for example, between two oppositions or two inferior conjunctions) depends on the relative motions of the Earth and the planet, and is called the **synodic period**. Synodic periods are greatest for the two planets that are closest to the Earth (Venus and Mars). They are like athletes running in adjacent lanes on a circular track at speeds that are not greatly different; it takes a long time for the athlete on the inner lane to lap the one on the outer lane. For example, the sidereal period of Mars is 687 days (1.88 years) and that of the Earth, 365 days (1 year). After one year, the Earth has made one complete journey around the Sun, whereas Mars has travelled only just over halfway round its orbit. Because the Earth still has to continue the chase for a long time before catching up again, the average time interval between two successive oppositions of Mars is 780 days (2.14 years). By way of contrast, the Earth overtakes slow-moving Saturn (which has an orbital period of 29.46 years) at intervals of one year and thirteen days.

The phases of the Moon

The Moon shines because it reflects sunlight. At any instant, one hemisphere is illuminated by the Sun and the other is in darkness. Consequently, as the Moon moves around its orbit, and the angle between the Sun and Moon in the sky changes, the proportion of the Earth-facing hemisphere that is illuminated increases and decreases, and the apparent shape – or **phase** – of the illuminated face of the Moon varies in a regular cycle (Fig. 3.5).

At '**New Moon**' the Sun and the Moon are close together in the sky and the Moon's Earth-facing side is in darkness. Thereafter, the Moon moves steadily to the east of the Sun at a rate of about 12° per day. Its phase increases from a thin crescent, lit up on its west-facing edge, to a half-illuminated disc in about a week; this 'half-moon' phase, which

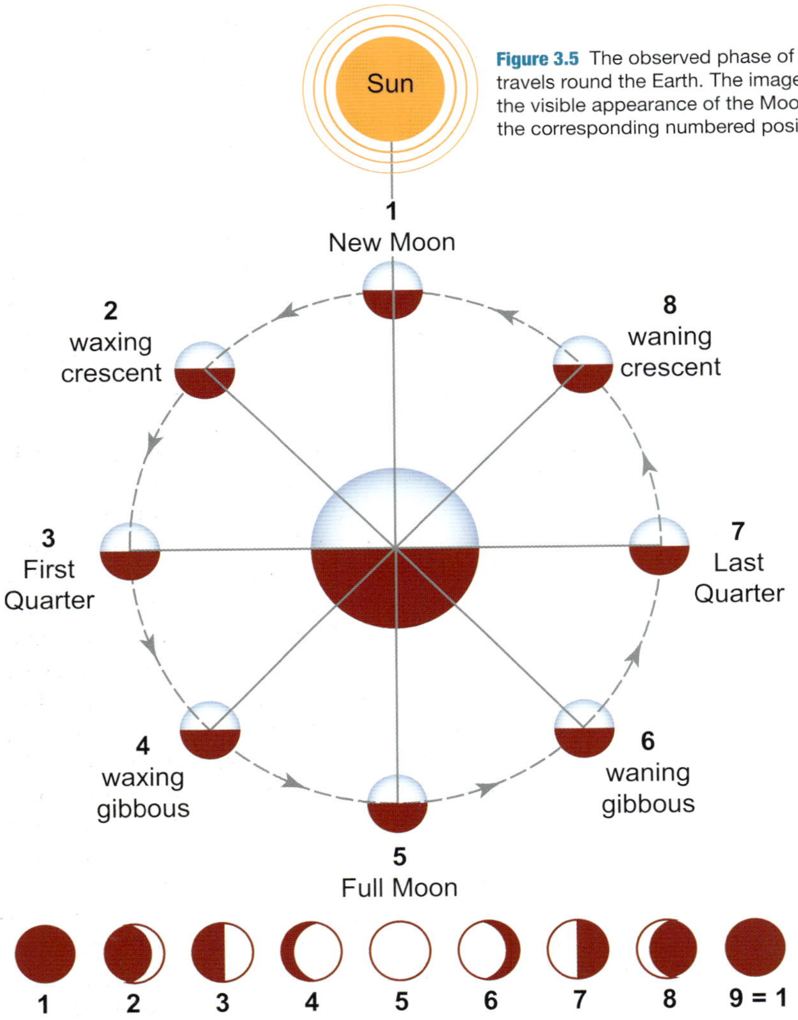

Figure 3.5 The observed phase of the Moon changes as it travels round the Earth. The images along the bottom show the visible appearance of the Moon when the Moon is at the corresponding numbered positions on its orbit.

is called **first quarter**, occurs when the Moon has travelled one quarter of the way round the Earth. The angle between the Sun and the Moon in the sky (the Moon's elongation) is then 90°, and the Moon sets approximately six hours after the Sun. The phase then becomes gibbous (greater than half) and continues to grow until, after a further week, the Moon's disc is fully illuminated (**Full Moon**); the Moon is then on the opposite side of the Earth from the Sun, rising in the east at around sunset, culminating at midnight, and setting in the west about the time of sunrise. Thereafter, the Moon approaches closer to the Sun in the sky and its phase diminishes, reaching **last quarter** ('half moon') about a week after Full, then shrinking to a thin crescent, which rises just before dawn, before returning to New Moon.

The Moon's sidereal period (the time it takes to travel once around the Earth relative to the background stars) is 27.32 days. If the Earth were stationary, the time interval between two successive New Moons (the Moon's synodic period) would be identical to its sidereal period. However, in the time that it takes for the Moon to complete one orbit round the Earth, the Earth itself has travelled through an angle of about 27° along its orbit round the Sun, so that after 27.32 days, the Sun and the Moon are not yet back in line with each other. Because the Moon has to travel along its orbit for a further two days before New Moon recurs, its synodic period (which otherwise is known as the lunar month) is 29.53 days.

Eclipses of the Sun and Moon

An **eclipse** of the Sun occurs when the Moon passes directly in front of the Sun and wholly or partly hides it from view. An eclipse of the Moon takes place when the Moon passes wholly or partly into the shadow cast by the Earth. If its orbit lay exactly in the same plane as that of the Earth, the Moon would pass directly between the Earth and the Sun at each New Moon and through the Earth's shadow at each and every Full Moon. In fact, because the Moon's orbit is inclined to that of the Earth by an angle of about 5°, the Moon usually passes a little way above or below the Sun at New Moon, and a little way above or below the Earth's shadow at Full Moon. An eclipse can take place only if New Moon or Full Moon occurs when the Moon is at, or relatively close to, one of the points (nodes) at which the Moon's orbit crosses the plane of the Earth's orbit.

The shadow cast by the Moon consists of two parts, a cone of dark shadow called the **umbra**, and an outer region called the **penumbra**. For an observer located inside the umbra, the sun will be completely hidden and a **total eclipse** of the Sun will be observed. Within the penumbra, the Sun is only partly hidden, and a **partial eclipse** is observed (Fig. 3.6).

Although the diameter of the Sun is nearly 400 times greater than that of the Moon, the Sun is also about 400 times further away. A consequence of this curious quirk of nature is that the Sun and the Moon look almost exactly the same size in our skies. Because the Moon travels round the Earth in an elliptical orbit, its distance varies, and it therefore appears smaller at **apogee** (its most distant point) than at **perigee** (its point of closest approach); likewise, because the Earth travels round the Sun in an elliptical path, the apparent size of the Sun varies, too. As a result, the Moon sometimes appears smaller than the Sun and its cone of dark shadow does not quite reach as far as the Earth. In these circumstances, if the Moon passes directly between the Sun and the Earth, it appears as a dark disc surrounded by a narrow ring, or annulus, of sunlight; such a phenomenon is called an **annular eclipse**. Even when the Moon's umbra does reach as far as the Earth, it is always very narrow (never more than 270 km wide), and so a total solar eclipse can be seen only from a narrow strip of the Earth's surface as the Moon's shadow sweeps by. By contrast, a partial eclipse may

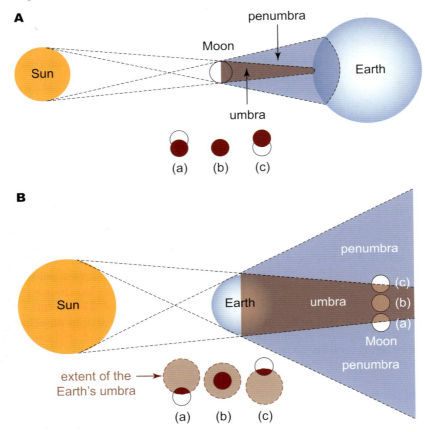

Figure 3.6 Eclipses: A, an eclipse of the Sun occurs when the Moon's shadow falls on the Earth. An observer within the umbra will see a total eclipse (b); an observer in the penumbra will see a partial eclipse (a) or (c). B, a lunar eclipse occurs when the Moon enters the Earth's shadow. If the whole of the Moon enters the umbra (b) a total eclipse of the Moon occurs; if part of the Moon enters the umbra (a) or (c), a partial eclipse is seen.

be seen from a substantial fraction of the daytime hemisphere. Although total eclipses are not particularly rare in themselves, the narrowness of the cone of shadow ensures that they are only seen very occasionally from any particular location.

Because the Earth is about four times larger than the Moon, its shadow is wider and extends further. If the Moon passes completely into the Earth's umbral shadow, a total lunar eclipse will occur; if only part of the Moon's disc enters that shadow, a partial eclipse will be seen.

A total eclipse of the Sun is one of the most dramatic of natural phenomena. It is little wonder that ancient peoples were terrified

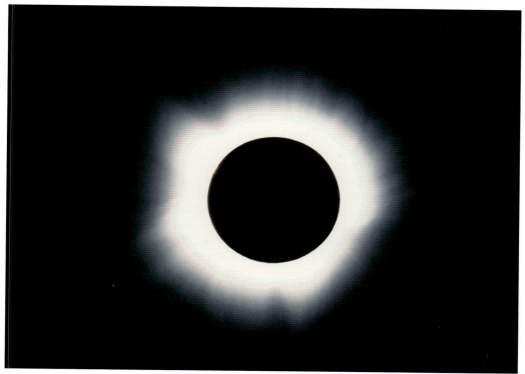

Figure 3.7 This image of the total solar eclipse of 7 March 1970 shows the delicately structured glow of the solar corona beyond the dark disc of the Moon. Image credit: NSO/AURA/NSF.

by them, and that nowadays thousands of people are prepared to travel to the ends of the Earth to witness these awe-inspiring events (Fig. 3.7). But eclipses of any kind – whether solar or lunar – are fascinating phenomena that testify to the interplay between the motions of Sun, Earth and Moon.

Gravity and orbits

According to Sir Isaac Newton's law of universal **gravitation**, which he published in 1687 in his celebrated work *Philosophiae naturalis principia mathematica* ('The mathematical principles of natural philosophy'), each body attracts every other one with a force that depends on the masses of, and separations between, those bodies. The gravitational force of attraction between two bodies, of masses m and M, depends on the product of their masses (m × M) and decreases in proportion to the square of the distance between them. If the distance is doubled, the force of attraction is reduced to a quarter of its previous value ($\frac{1}{2}^2 = \frac{1}{4}$). Subject to an 'inverse square' force of this kind, planets move round the Sun in accordance with Kepler's laws.

According to Newton's first law of motion, a body continues to move in a straight line at a constant speed unless acted on by a force. If a body were thrown parallel to the ground from the top of a tall tower, and if the Earth's gravity could somehow be 'switched off', then, neglecting the effects of air resistance, that body would continue to move in its original direction at a constant speed. In reality, gravity will accelerate the body towards the centre of the Earth, and the body will hit the Earth's surface some distance away from the base of the tower. The faster it is thrown, the further it will travel before striking the ground. If the body is thrown at precisely the right speed, parallel to the Earth's surface, the combination of transverse (sideways) motion and radial (downwards) motion will cause it to move round the Earth in a circular path. This velocity (velocity = speed in a particular direction) is called **circular velocity** (Fig. 3.8).

Close to the Earth's surface, circular velocity has a value of 7.8 km/s (about 28,000 km/h), but it decreases with increasing distance. At a distance of 42,000 km from the centre of the Earth, a satellite in a circular orbit travels at 2.9 km/s and takes 24 hours to travel all the way around our planet. If such a satellite is orbiting directly above the Equator, it will remain permanently above a particular point on the Earth's surface and will appear to be stationary in the sky; this **geostationary orbit** is used extensively by communications satellites. The much more distant Moon travels at about 1 km/s and takes 27.32 days to travel round our planet. The Moon also rotates around its axis in precisely the same period of time. Consequently, it keeps the same hemisphere turned towards the Earth at all times; the other hemisphere, which always faces away from the Earth, remained unseen until 1959, when the Soviet spacecraft Luna 2 flew

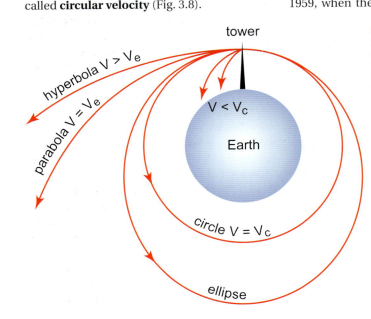

tower

hyperbola V > Ve

parabola V = Ve

$V < V_c$

Earth

circle V = Vc

ellipse

Figure 3.8 If a projectile were fired parallel to the Earth's surface from a very tall tower at less than circular velocity (Vc), it would fall to the ground; if fired at circular velocity it would enter a circular orbit; if fired at a speed greater than circular velocity but less than escape velocity (Ve), it would enter an elliptical orbit. If fired at, or in excess of, escape velocity, it will escape along a parabolic or hyperbolic path.

past the Moon and sent back the first grainy images of its far side.

If a projectile is thrown upwards at a modest speed, it will reach a maximum height and then fall back to the ground. If it is thrown faster, it will reach a greater height before falling back, and if it is fired at a high enough speed, it will continue to recede forever. The minimum speed at which a projectile must be fired in order never to fall back is called **escape velocity**. Escape velocity at the surface of the Earth is 11.2 km/s (about 40,000 km/h), but its value decreases at greater distances.

If the speed of an orbiting spacecraft is increased beyond circular velocity, it will enter an elliptical orbit, the point of closest approach to the Earth being called **perigee**, and the point at which its distance is greatest, **apogee**. The greater the impulse (increase in speed), the more elongated the resulting ellipse becomes. If the speed is increased to the precise value of escape velocity at that point, the spacecraft will move away along a parabolic path, and will never return. Its speed will decrease as it recedes, approaching ever closer to, but never quite reaching, zero. If the speed of the spacecraft exceeds escape velocity, it will continue to recede at a finite speed forever.

In order to leave the Earth's neighbourhood and travel off into interplanetary space, a spacecraft must exceed the Earth's escape velocity. By the time the spacecraft has receded to a substantial distance, the Earth's gravitational attraction will be small compared to that of the Sun, and the spacecraft's trajectory will be controlled thereafter by the Sun's gravitational field; it will then behave like a tiny planet orbiting the Sun. In terms of energy requirements, the most economical way to send a spacecraft from the Earth to Mars, Jupiter or beyond is to launch it in the same direction as the Earth itself is moving, so that the resulting speed of the spacecraft relative to the Sun will be equal to the final speed of the spacecraft relative to the Earth plus the orbital speed of the Earth itself. The spacecraft then enters an elliptical orbit with perihelion equal to the Earth's distance, and aphelion equal to the distance of the target planet. To reach a planet that is closer to the Sun (Mercury or Venus), the spacecraft needs to be fired in the opposite direction to the Earth's orbital motion, so that it ends up moving more slowly than our planet and enters an elliptical orbit that drops in closer to the Sun.

The disadvantage of orbits of this kind is that flight times are long (about nine months in the case of a mission to Mars, for example). Faster routes are possible, but they require more energy and greater expenditure of fuel. One way of getting round this problem is to make use of '**gravitational slingshots**'. When a spacecraft approaches another planet and falls under its gravitational influence, it accelerates, loops round the planet in a hyperbolic path, and then recedes in a different direction. Its final speed relative to the planet will be the same as its initial speed of approach, but, so far as the Sun is concerned, its speed will then be equal to the speed at which the spacecraft is receding from the planet plus the speed at which the planet itself is moving (rather like jumping onto a moving roundabout then leaping off in the direction in which the roundabout is turning). By this means, very substantial changes in velocity can be achieved.

In some cases, several different planetary encounters and slingshots can be, and have been, used to send spacecraft to their ultimate

A

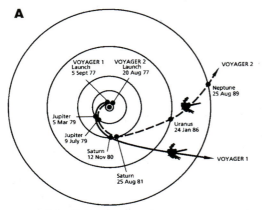

VOYAGER 1
Launch
5 Sept 77

VOYAGER 2
Launch
20 Aug 77

VOYAGER 2

Neptune
25 Aug 89

Jupiter
5 Mar 79

Uranus
24 Jan 86

Jupiter
9 July 79

Saturn
12 Nov 80

VOYAGER 1

Saturn
25 Aug 81

Figure 3.9 **A,** The trajectories that enabled NASA's twin Voyager spacecraft to tour the four giant planets and achieve high enough velocities to escape from the Solar System. Image: NASA. **B,** The journey of ESA's Rosetta spacecraft from launch in 2004 to its encounter with comet 67P/Churyumov-Gerasimenko in August 2014 included three fly-bys of Earth and one of Mars. Image: ESA-C. Carreau.

B

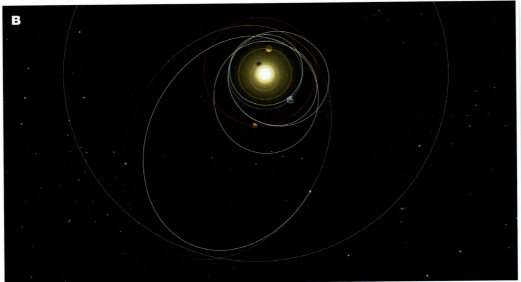

destinations (Fig. 3.9). For example, successive gravitational slingshots at Jupiter (in 1979) and Saturn (1980) accelerated the Voyager 1 spacecraft to a speed in excess of the escape velocity of the Solar System; heading towards interstellar space, it passed beyond the **heliopause**, the 'official' boundary of the Solar System (*see* Chapter 4), in August 2012. In contrast, the MESSENGER spacecraft, launched in 2004, made use of one flyby of the Earth, two of Venus and three of Mercury to ensure that on its next encounter with that planet, in 2011, it was approaching sufficiently slowly to be placed into a closed orbit around it.

4   The Sun

The Sun, our neighbourhood star, is a self-luminous gaseous body. A modest and unexceptional star, it nevertheless is by far the most massive body in the Solar System. Its diameter of 1,392,000 km is 109 times greater than that of the Earth but, although its vast globe could contain 1.3 million bodies the size of our planet, its mass is only about 330,000 Earth masses, and its mean density, therefore, is about a quarter of the mean density of our rocky planet. This relatively low density (1,408 kg/m³), together with data gathered from spectroscopic observations (*see* Chapter 6), indicates that the Sun is composed primarily of hydrogen (73.5%) and helium (25%), with heavier elements making up only about 1.5% of the total.

Being a gaseous body, the Sun does not have a solid surface. Its visible surface, which is called the **photosphere** (literally, 'sphere of light'), is the layer from which the Sun's light escapes into space. Below the photosphere, the Sun's material is opaque, whereas above the photosphere, the rest of the solar atmosphere is, in essence, transparent to visible light.

The Sun's effective surface temperature is 5780 K (about 5500 °C) and its **luminosity** (the total amount of energy that it radiates into space, every second, from its entire surface), is 3.86×10^{26} watts (W); this is equivalent to the power that would be radiated by 386,000 billion, billion 1-kilowatt heaters. The energy radiated by the Sun spreads out equally in all directions and the **flux** (the amount of energy per second passing perpendicularly through an area of one square metre), is inversely proportional to the square of distance (if the distance is doubled, the flux is reduced to one quarter of its previous value). By the time it arrives at the top of the Earth's atmosphere, the solar energy flux is reduced to 1.37 kW per square metre (kW/m²), a quantity that is called the **solar constant** (in reality, the amount of energy reaching the Earth varies slightly because the distance between the Earth and the Sun alters as the Earth moves along its elliptical orbit, and because of minor fluctuations in the Sun's output).

The Sun's energy source

The Sun generates and sustains its prodigious output of energy by means of nuclear **fusion** reactions that take place deep down in its central core, where the temperature is about 15,000,000 K and solar material, squeezed together by the immense weight of the overlying layers, is 160 times denser than water. Under these extreme conditions, solar material is fully ionized (i.e. its atoms have been stripped of their orbiting electrons), and nuclei of hydrogen atoms (the **nucleus** of a hydrogen atom, chemical symbol H, consists of a single, positively charged, **proton**) collide with sufficient violence to fuse together to form nuclei of the second-lightest element, helium (He). The principal fusion reaction that powers the Sun is known as the **proton–proton reaction**. It is a three-stage process that requires the participation of a total of six protons (hydrogen nuclei), two of which are converted into **neutron**s and two of which are set free at the final step. The net result is that four hydrogen nuclei are welded together to make one helium nucleus, which contains two protons and two neutrons.

The mass of the resulting helium nucleus is about 0.7% less than the combined mass of the particles that went into its formation. The mass that is 'lost' is liberated in the form of energy, in accordance with Einstein's well-known relationship, $E = mc^2$, whereby, if a quantity of matter is converted into energy, the energy (E) released is equal to the mass (m) that is 'destroyed' multiplied by the square of the **speed of light** (which is denoted by the symbol, c). Because the speed of light is a very large number (about 3×10^8 m/s), the speed of light squared is an exceedingly large number, and the amount of energy liberated by the 'destruction' of a small amount of matter is extremely large. For example, if one kilogram of anything (for example, copies of this book) could be turned completely into energy, the energy released would be about 9×10^{16} joules (J) ($E = mc^2 = m \times c \times c = 1 \times (3 \times 10^8) \times (3 \times 10^8) = 9 \times 10^{16}$ J) – enough to power a 1-kilowatt heater for nearly three million years.

In every second, the Sun converts some 600 million tonnes of hydrogen into helium and effectively destroys (by converting it into energy) about 4.4 million tonnes of matter. Gradually, over long eons of time, the Sun's core is being converted from hydrogen into helium. Astronomers believe that the Sun has been shining in this way for about 4.6 billion years, and that it has enough fuel to sustain it for a further 5 billion years or so; there is no danger of the Sun running out of fuel any time soon!

The structure of the Sun

The solar **core**, where energy is generated by fusion reactions, extends out to about one quarter of the Sun's radius. Surrounding the core, and stretching out to about 70% of the solar radius, is a region that is called the **radiative zone**. In this layer, energetic **photons** ('particles' of electromagnetic radiation) flow outwards from the core towards the surface. In the dense ionized gas that fills the solar interior, a photon can travel only a tiny distance before colliding with an electron or atomic nucleus and bouncing off (scattering) – or being absorbed and re-emitted – in a random direction. Because of the vast number of collisions it experiences, a photon can easily take tens, or even hundreds of thousands of years, to make its way from the core to the surface of the Sun. The radiative zone is surrounded by a layer of cooler gas, called the **convective zone**, within which energy is transported by currents of hot gas that rise towards the photosphere, give out heat, and then sink back down to be heated once again by energy flowing from the underlying radiative zone (Fig. 4.1).

Much of the Sun's energy escapes in the form of visible light from the photosphere, the thin layer within which the solar gas changes from being opaque to transparent, and it then flows largely unimpeded out into space. A photon that has taken, perhaps, a hundred thousand years to battle its way from the core to the surface, can then travel across the intervening 150 million kilometres of space to reach the Earth just over eight minutes later

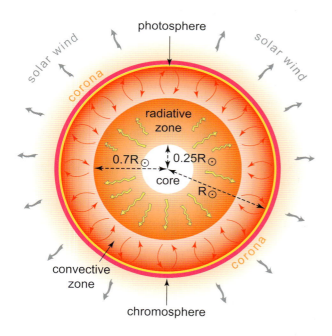

Figure 4.1 Cross-section of the Sun: energy from the core is transported through the radiative zone by photons, then by convection through the convective zone to the photosphere (the visible surface), beyond which lies the chromosphere, and the corona, out of which the solar wind flows.

(there to be absorbed, perhaps, by a sunbather lying on a beach!)

The photosphere is only about 500 km thick. When we look directly down onto the centre of the Sun's visible disc (*warning: on no account look directly at the Sun with a telescope, binoculars or even the naked eye – to do so would be to risk serious eye damage, or permanent blindness*), our line of sight penetrates to deeper levels of the photosphere, where the temperature is higher, and the Sun's light is more brilliant. When we look towards the edge of the Sun's disc (which astronomers call the solar '**limb**'), our line of sight penetrates at a shallow angle into the cooler outer levels of the photosphere, where sunlight is less intensely brilliant. For that reason, the Sun exhibits a phenomenon called **limb**

darkening, whereby the region around the edge of the solar disc appears less bright than its centre.

The thin, and much more tenuous, atmospheric layer – some 2000–10,000 km thick – that lies immediately above the photosphere is called the **chromosphere** (literally, 'colour sphere') because of the pink-red hue that it displays when the dazzling photosphere is blotted out by the Moon during a total solar eclipse. Under these circumstances, the chromosphere appears as a narrow ring around the dark disc of the Moon (its colour comes from red light emitted by hydrogen atoms in that tenuous gaseous layer). Much too faint to be seen with the naked eye except during a total eclipse, the chromosphere can, however, be studied at any time with specialized instrumentation. Beyond the chromosphere, and extending out to several times the radius of the main body of the Sun, is the Sun's outer atmosphere – the **corona**.

The corona is exceedingly faint. At visible wavelengths it has only about one millionth of the brilliance of the photosphere, and so can only be seen with the naked eye at a total solar eclipse; however, it can be studied at any time by specialized instruments called coronagraphs, which produce artificial eclipses, and which are carried on satellites and spacecraft clear of the obscuring influence of the atmosphere and the scattered sunlight that hides the corona from our eyes in the daytime sky. The corona is a **plasma** – a mixture of positively and negatively charged particles, predominantly protons (hydrogen nuclei) and electrons. The temperature of the corona – where temperature in this context is a measure of the energy of its constituent particles, which are rushing about, and colliding with each other, at very high speeds – is well in excess of one million kelvins (K), far higher than the temperature of the photosphere. It seems so much fainter than the photosphere, at visible wavelengths, simply because it is so tenuous.

Overall, the temperature of the solar atmosphere decreases from about 6400 K at the base of the photosphere to about 4200 K at the base of the chromosphere. Thereafter, it begins to increase, gently at first, but then shoots up very rapidly from about 10,000 K at the top of the chromosphere, to more than 1,000,000 K in the corona itself. The corona is heated by processes that take place in magnetic fields that are embedded within it and, because of its very high temperature, it radiates most strongly at ultraviolet and x-ray wavelengths.

Solar activity and the role of magnetic fields

The most obvious features on the surface of the Sun are **sunspots** – dark patches on the photosphere that range in size from tiny spots (called pores) at the limit of visibility to huge groups that cover billions of square kilometres (Fig. 4.2). Sunspots are cooler than the photosphere as a whole, and so appear dark by contrast with their surroundings. A substantial spot will normally consist of a darker central region called the **umbra**, where the temperature may be as low as 4000 K, surrounded by a less dark **penumbra**, where the temperature is about 5500 K; the ambient temperature of the photosphere is about 6000 K.

Sunspots are associated with areas of concentrated, localized **magnetic fields** on, and immediately below, the solar surface. The magnetic field in a typical spot is about ten thousand times stronger than the magnetic field (which causes compass needles to point towards the terrestrial magnetic poles) at the surface of the Earth. Strong magnetic fields inhibit the process of convection that brings heated gas to the surface of the Sun, and so these regions radiate less heat, and are cooler and darker than, their surroundings. Sunspots usually (but not always) appear in pairs (or more complex groups), a typical pair comprising one spot with north magnetic polarity and a neighbour with the opposite (south) polarity. Spot pairs behave as if there were a bar magnet embedded beneath the region of the photosphere where the spot pair appears. Magnetic lines of force extend outwards from the north polarity spot, loop round and

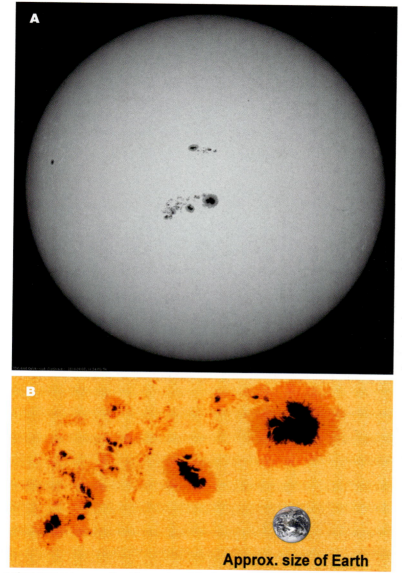

Figure 4.2 **A**, a giant sunspot group sits at the centre of the Sun on 7 January 2014; the image also shows how the brightness of the photosphere declines towards the edge of the disc (limb darkening). **B**, a close-up view of the same sunspot group, showing the umbra and penumbra of the various spots, and indicating the size of the Earth to scale. Image credit: NASA/SDO.

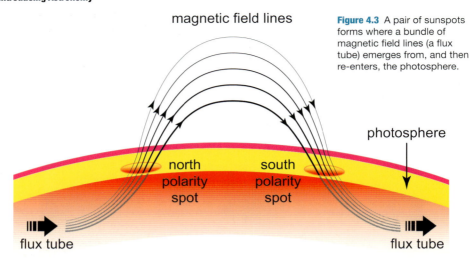

magnetic field lines

Figure 4.3 A pair of sunspots forms where a bundle of magnetic field lines (a flux tube) emerges from, and then re-enters, the photosphere.

photosphere

north polarity spot

south polarity spot

flux tube

flux tube

re-enter the photosphere at the adjacent, south polarity, spot. Groups consist of more complex regions of opposing magnetic polarity. A pair of sunspots, and the associated bipolar magnetic region (or '**active region**'), is created when a bunch of magnetic field lines (a 'flux tube') develops a kink and penetrates the photosphere (Fig. 4.3).

Sunspots form, develop, decline and disappear over periods of time ranging from less than a day to several weeks, or even months. Sunspots, and groups, share in the rotation of the Sun, and the first measurements of solar rotation were made by studying the rates at which the spots tracked across the solar disc. Observations of this kind revealed that the Sun does not rotate like a solid body (where every point on the surface – being rigidly fixed relative to its neighbours – rotates round the central axis in the same period of time) but instead exhibits **differential rotation**, the rotation period close to the solar equator being just under 25 days; the surface

rotation period increases with increasing latitude, being about 27 days at latitudes 30° north or south of the solar equator, and about 34 days close to the poles.

Prominences, flares and coronal mass ejections

When the sun is viewed through filters that transmit light of one particular wavelength only (for example, one of the wavelengths at which hydrogen emits, or absorbs, light), most of the background brilliance of the photosphere is cut out and the resulting image reveals structures in the chromosphere and corona. Observations of this kind reveal the presence of bright **prominences** – glowing clouds of gas which look almost like flames, projecting beyond the visible limb of the Sun (Fig. 4.4). These same features, when viewed against the bright background of the solar disc, appear as dark **filaments**, which are absorbing light coming from the photosphere below. Prominences and filaments are clouds of gas that are denser than their surroundings,

Figure 4.4 A huge handle-shaped prominence is shown (upper right) in this ultraviolet image obtained by the NASA/ESA SOHO spacecraft. The dark, elongated features across the solar disc are filaments; the white patches are active regions where magnetic fields are concentrated. Image credit: SOHO (ESA & NASA).

and are suspended in the solar atmosphere by magnetic fields. There are two basic types of prominence – quiescent and eruptive (or active). Quiescent prominences may hang like clouds in the solar atmosphere for weeks, or even months, with little overall change. In contrast, eruptive prominences surge up and down on short timescales, sometimes reaching heights of several hundred thousand kilometres, or even ejecting material right out into interplanetary space.

Solar flares are explosive outbursts in which up to 10^{25} J (equivalent to the energy released by several billion 1 megaton nuclear bombs) can be released within a few minutes (Fig. 4.5). Flares radiate energy over a very wide range of the **electromagnetic spectrum** from gamma rays to radio waves, the bulk of their energy being released in the form of x-rays and extreme ultraviolet radiation.

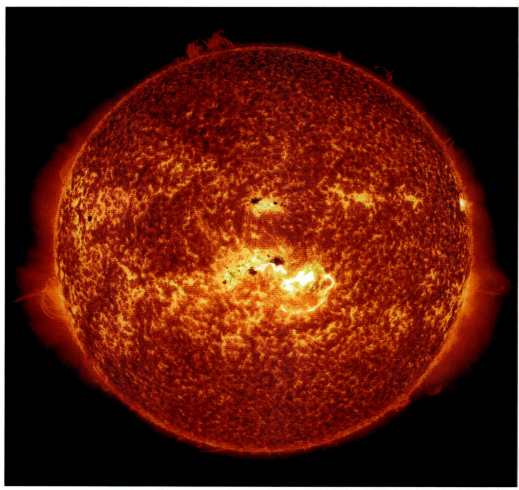

Figure 4.5 The very bright patch just below right of centre is a major flare that erupted from the active region associated with the large sunspot group featured in Figure 4.2; imaged by NASA's Solar Dynamics Observatory spacecraft. Image credit: NASA/SDO/Goddard Space Flight Center.

They eject streams of charged particles, including electrons that have been accelerated to speeds of up to half the speed of light, and atomic nuclei (mainly hydrogen nuclei), and they catapult clouds of plasma out through the corona. Flares are caused by the sudden release of magnetic energy that has been stored in the twisted magnetic fields associated with complex active regions. The process that triggers these events is called **magnetic reconnection**: when oppositely-directed magnetic field lines come into contact with each other, they join up to form new loop-like structures with their feet embedded in or near sunspot groups. Like an elastic string, a magnetic field line has tension in it; when it severs and reconnects it catapults great blobs of plasma through the corona out into interplanetary space. Huge bubbles of plasma containing billions of tonnes of material, which erupt forth from the corona and propagate through interplanetary space, are known as **coronal mass ejections** (CMEs) (Fig. 4.6).

Figure 4.6 A huge coronal mass ejection (CME) surges outwards (upper left) in this composite image recorded by the SOHO spacecraft on 4 January 2002. Image credit: SOHO (ESA & NASA).

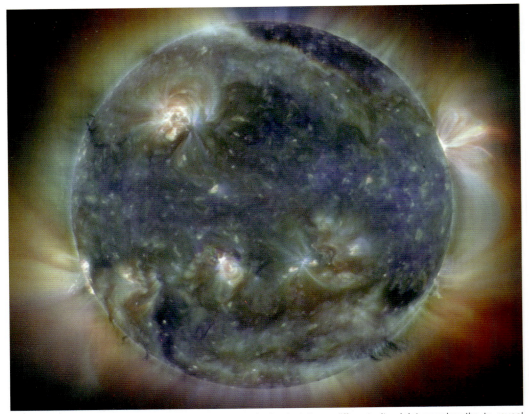

Figure 4.7 This colour composite combines images obtained at three different ultraviolet wavelengths to reveal structures in the inner part of the corona (active regions and magnetic loops) together with a large, darker, coronal hole (at the top). Image credit: ESA/NASA/SOHO.

X-ray and extreme ultraviolet images of the Sun show that the corona is uneven and clumpy in structure. It contains bright 'active' regions where concentrated hot plasma is confined by loops of magnetic field lines, quiet regions of lesser emission, and apparently dark, low-density regions called **coronal holes**, where magnetic field lines trail out into interplanetary space, thereby enabling solar plasma to escape from the Sun (Fig. 4.7).

There is a continuous flow of charged particles (mainly electrons and protons) from the corona into interplanetary space. Known as the **solar wind**, this outflow blows past the planets at speeds in excess of a million kilometres per hour and causes the Sun to lose about a million tons of material every second. The outward flow of the solar wind carries with it lines of force of the solar magnetic field, which spread out to form the weak interplanetary magnetic field. The solar wind interacts with the magnetospheres of planets (a **magnetosphere** is the region of space that is occupied by a planet's magnetic field) and the tails of comets (*see* Chapter 5). Fluctuations caused by coronal mass ejections, or bursts of high-speed particles ejected by solar flares, affect planetary magnetospheres, compressing and stretching them, and sometimes severing the tails of comets. In a very real sense the Earth, and the rest of the Solar System, inhabits the outer regions of the solar atmosphere, and is directly affected by solar storms and fluctuating conditions in the space environment, which are called **space weather**.

The solar wind continues to flow outwards, becoming ever more thinly spread, until it is halted by the pressures exerted by the exceedingly tenuous gas that lies between the stars, and by the feeble magnetic fields that permeate **interstellar space**. The Sun's magnetic domain is called the **heliosphere**, and its boundary – which is called the **heliopause** – lies well beyond the orbit of the planet Neptune, its size varying with changing levels of solar activity. The Voyager 1 spacecraft, which was launched in 1977 and flew past Jupiter and Saturn before heading out towards interstellar space, is believed to have crossed the heliopause on 25 August 2012, by which time it had reached a distance of 121 AU (18 billion kilometres) from the Sun.

The solar cycle

All forms of **solar activity** – sunspots, prominences, flares, coronal mass ejections, together with the shape and structure of the corona and the overall magnetic field of the Sun – display cyclic variations with a period of about 11 years, a pattern of behaviour that is called the **solar cycle** (Fig. 4.8).

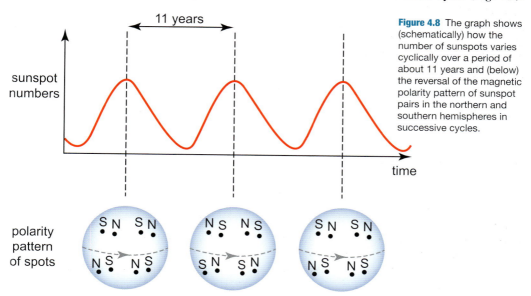

Figure 4.8 The graph shows (schematically) how the number of sunspots varies cyclically over a period of about 11 years and (below) the reversal of the magnetic polarity pattern of sunspot pairs in the northern and southern hemispheres in successive cycles.

The number of spots and groups reaches a maximum once every 11 years or so and thereafter, over the course of the following six or seven years, spot activity declines to a minimum. Around the time of solar minimum, the face of the Sun may be devoid of spots for days, or even weeks, on end; thereafter, over the course of the next four to five years, activity rises again, towards the next maximum. The level of activity can differ significantly between successive cycles, and there is evidence for longer-term modulation of sunspot numbers with cycles of 80 years or longer, and of occasional periods in the past (such as the Maunder minimum of 1645–1715) during which solar activity has almost completely vanished for periods of several decades.

The first spots of a new cycle usually appear in two latitude bands, about 30–40 degrees either side of the solar equator and, as the cycle advances, the bands of activity migrate towards the equator. The polarity pattern of sunspot pairs and groups reverses every 11 years (so if, in one cycle, the leading spot in each northern hemisphere spot pair has north magnetic polarity, in the subsequent cycle, the leading spot in northern hemisphere pairs will have south magnetic polarity), so that the overall magnetic cycle repeats at intervals of 22 years.

The solar magnetic field, which controls this cycle, is believed to be generated and sustained by circulating currents in the convective zone. Differential rotation causes material close to the equator to pull ahead of material at higher latitudes, and so distorts and stretches the embedded magnetic field lines, amplifying the strength of the field where the lines become bunched together.

When the magnetic field contained in a bundle of field lines (a '**flux tube**') becomes sufficiently strong, the flux tube floats to the surface and erupts through the photosphere to form an active region and an associated pair of spots. The winding action of differential rotation gradually drags the magnetically active band closer to the equator. The magnetic polarity associated with the following spot (the member of the pair which is behind its neighbour, in the sense of the Sun's rotation) migrates preferentially towards the rotational pole and, over the course of 11 years or so, the cumulative effect of this process reverses the overall magnetic polarity of the Sun.

The solar cycle has direct effects here on Earth. Solar outbursts, coronal mass ejections, and fluctuations in the solar wind compress the Earth's magnetosphere, causing electrical current to flow in the upper atmosphere and producing disturbances, known as **magnetic storms**, which affect compass needles and cause surges in power transmission lines, which, on occasion, can lead to major power outages and blackouts. At times of high solar activity, when flares are more prevalent, the output of x-ray and ultraviolet radiation from the Sun can be greatly enhanced; the resultant effects on the **ionosphere** (*see* Chapter 5) can have a major impact on radio communication. Impacting bursts of energetic solar particles causes changes in the structure of the Earth's magnetosphere that catapult charged particles down into the upper atmosphere, where they interact with atoms and ions, causing them to emit light; this creates the shimmering, shifting patterns of light – often seen in polar regions – which are called the **aurorae** (Fig. 4.9). The most energetic bursts of solar

Figure 4.9 A bright aurora illuminating the sky near Tromso, Norway, on 17 February 2013. Image credit: ESA-S. Mazrouei.

particles can damage satellites and cause computers to crash.

With human activity being so heavily dependent on electrical power and on electronic devices and communication systems, it is little wonder that very close attention is now paid to monitoring solar activity and space weather.

5 The Sun's domain

The Sun lies at the heart of the Solar System. Of its eight orbiting planets, the innermost four – Mercury, Venus, the Earth and Mars – are small, dense rocky worlds, broadly similar in general nature to the Earth, and known, therefore, as the **terrestrial planets**. The more distant four – Jupiter, Saturn, Uranus and Neptune – are called the **giant planets**, or **gas giants**, because they are much larger than the Earth and are composed primarily of hydrogen, helium and hydrogen compounds, such as methane and ammonia.

The Earth

The Earth is the largest and most massive of the terrestrial planets. It is unique among the known planets in having liquid water at its surface (oceans cover about 70% of its surface), an atmosphere that contains large amounts of oxygen, and in supporting a diverse multitude of life-forms (Fig. 5.1).

The Earth has an outer **crust** consisting of relatively low-density rocks, which is about 35 km thick, on average, in the continental land masses, but only some 7 km thick in the

Figure 5.1 The Earth as seen from the Apollo 17 spacecraft in 1972. The vista extends from the Mediterranean to the icy continent of Antarctica and includes the Atlantic and Indian Oceans, together with the storm clouds of the Southern Ocean. Image credit: NASA.

ocean beds. The crust lies on top of a **mantle** of denser rock that extends down to a depth of about 2900 km. The **core**, which is composed predominantly of iron, together with smaller quantities of nickel, has a radius of about 3500 km and a central temperature of about 6000 K. The outer part of the core is liquid, but the central part, which is under higher pressure, is solid (Fig. 5.2).

The outermost level of the mantle, together with the crust, forms a layer that is called the **lithosphere**. According to the theory of **plate tectonics**, the lithosphere has fractured into eight major pieces (called **lithospheric plates**) and a number of smaller ones, which sit on top of a warmer, less rigid, layer called the asthenosphere, which is sufficiently plastic to enable its constituent rocky material to flow, very slowly, in giant convection cells. Floating on top of this slow circulation, the plates are carried away from, or towards, each other at speeds of a few centimetres per year. At boundaries where two plates are moving apart, molten material may rise through the resulting gaps to create structures such as the Mid-Atlantic ridge, and where two plates collide and buckle, mountain ranges such as the Himalayas are thrust up. At boundaries where oceanic plates meet, and sink beneath, continental ones, mountain chains (for example, the South American Andes), strings of volcanoes, and deep ocean trenches are created.

The Earth's atmosphere is composed primarily of nitrogen (about 78% by weight) and oxygen (21%). Minor constituents include argon, carbon dioxide and water vapour. Because carbon dioxide and water vapour are efficient absorbers of infrared radiation, both of these gases are major contributors to the **greenhouse effect**. The basis of this effect is that incoming short-wavelength visible light from the Sun heats the ground, which then radiates longer-than-visible infrared radiation. A significant proportion of the outgoing infrared radiation is absorbed in the atmosphere, and then re-radiated back towards the ground, thereby raising the surface of our planet to an average temperature of around 290 K (17 °C), which is about 30 degrees higher than it would be if there were no atmosphere present.

The lowest level of the atmosphere – the **troposphere** – which extends to a height of about 10–12 km and is where most clouds and terrestrial weather systems occur, is heated primarily by energy radiated from the Earth's surface. For that reason, its temperature decreases with increasing altitude.

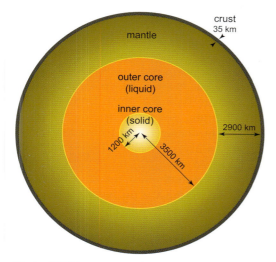

Figure 5.2 This cross-section shows the principal regions of the Earth's interior: the iron-nickel core (which comprises a solid inner core and a liquid outer core), the dense rocky mantle and the thin rocky crust.

Thereafter, the temperature begins to rise again, within the **stratosphere**, a layer that is heated primarily because it contains ozone (a molecule consisting of three oxygen atoms), which absorbs incoming ultraviolet light from the Sun. In the next layer – the **mesosphere** – the temperature declines again, before starting to rise once more in the highly tenuous **thermosphere**, a layer within which energetic solar radiation knocks electrons out of atoms and molecules to create the layers of **ions** and electrons that comprise the **ionosphere**.

The Earth has an overall magnetic field that behaves in some respects as if there were a bar magnet located deep inside the planet. The lines of force of this magnetic field emerge in the neighbourhood of the Earth's north magnetic pole, and re-enter around the south magnetic pole. The magnetic poles do not coincide with the rotational poles because, at present, the axis of the magnetic field is tilted to the rotational axis by an angle of about 11°. The magnetic field is generated and sustained by circulating currents in the rotating, electrically-conductive, liquid iron outer core. The region around the Earth within which the Earth's magnetic field is dominant is called the **magnetosphere**, and its boundary, the **magnetopause**. Pressure exerted by the solar wind and the interplanetary magnetic field squeezes the magnetopause inwards on the Sun-facing side, and draws it into a tail (the **magnetotail**) on the opposite, downstream, side.

The Moon

The Earth has one natural **satellite** – the Moon. The most obvious features of its surface are the light-coloured cratered highlands and the dark, comparatively smooth and lightly cratered plains – the lunar **maria**. The term 'maria' (singular **mare**), the Latin word for 'seas', reflects the fact that the early telescopic astronomers thought – incorrectly – that the dark plains were indeed seas and oceans and gave them names such as Mare Tranquillitatis (Sea of Tranquillity) and Oceanus Procellarum (Ocean of Storms), which are still used on lunar maps today. In reality, there is no liquid water anywhere on the Moon's airless surface, where the temperature ranges between a daytime maximum of just over 400 K (around 130°C) to a night-time minimum of about 93 K (−180°C) (Fig. 5.3).

The lunar craters, which range in size from tiny pits to huge structures up to several hundred kilometres across, are believed to have been formed by impacting rocky bodies, most of which rained down onto the lunar surface during the first few hundred million years after the Moon was formed. The dark, flat-floored maria were created by giant impacts that excavated huge basins (**impact basins**),which subsequently filled with magma that welled up from beneath the surface. Typical of these features is the Mare Imbrium (Sea of Clouds), which is some 1300 km in diameter. Part of the rim of the original basin is still visible in the form of three long, curved mountain ranges – the Apennines (to the south-east), the Carpathians (to the south) and the Alps (to the east and north). The dark mare rocks are similar to terrestrial basalts (volcanic rock typical of the Earth's ocean beds), whereas the highland rocks are lighter in colour and less dense. Although the dark maria are conspicuous features of the Earth-facing hemisphere of the Moon, they are almost entirely absent from the far side.

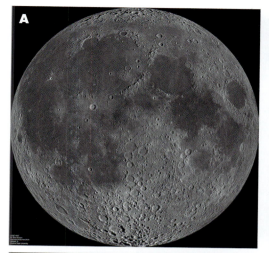

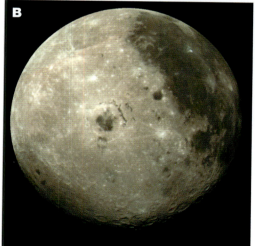

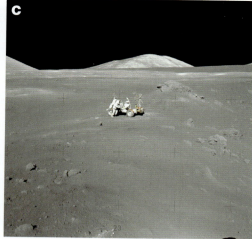

Figure 5.3 Vistas of the Moon: **A**, this mosaic of the Earth-facing hemisphere of the Moon, compiled from images obtained by the Lunar Reconnaissance Orbiter spacecraft, shows dark plains (maria), cratered highlands, and mountain chains, notably the Apennines and Alps on the edge of the Mare Imbrium (upper middle of image): NASA/GSFC/Arizona State University. **B**, with the exception of the small Mare Orientale, in the Orientale basin (centre), dark lava plains (maria) are almost completely absent from the far side of the Moon (left half of image). The large dark plain on the nearside (upper right) is the Oceanus Procellarum. Image credit: NASA/JPL. **C**, Lunar panorama showing Apollo 17 geologist-astronaut Harrison Schmitt beside the Lunar Roving Vehicle, with mountains and black sky beyond. Image credit: NASA.

Mercury

With a diameter of 4879 km, Mercury is the smallest of the four terrestrial worlds. Although it does not possess conspicuous magma-flooded basins like those that dominate the Earth-facing hemisphere of the Moon, its heavily cratered surface does display some particularly large impact structures, the biggest of which – the Caloris basin – has a diameter of some 1550 km (Fig. 5.4). Its surface temperature ranges from a maximum of 700 K (427°C) when the Sun is directly overhead at local noon, to a frigid 100 K (173°C) at local midnight. Intriguingly,

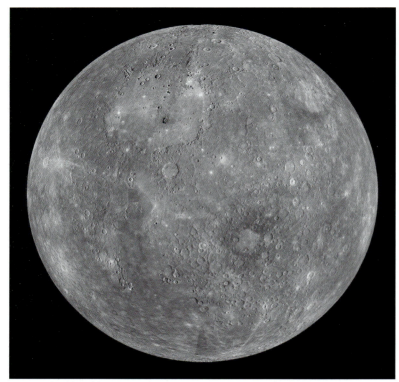

Figure 5.4 Mercury: this mosaic of images from the MESSENGER spacecraft shows the heavily-cratered surface of the planet and the conspicuous Caloris basin (upper left), which dominates the northern hemisphere. Image credit: NASA/Johns Hopkins University Applied Physics Laboratory/Carnegie Institution of Washington.

data from the MESSENGER spacecraft (which entered orbit round the planet in 2011) indicate that small quantities of water ice exist inside craters, close to the planet's north pole, which are in permanent shadow.

Despite its small size and mass, the average density of the planet is very nearly as great as that of the Earth. This implies that Mercury has a very large iron-nickel core, which contains about 70% of the planet's mass and extends out to about 75% of its radius (compared to 55% in the case of the Earth's core). Mercury has a weak, but significant, magnetic field, with about one

percent of the strength of Earth's. For several decades after the discovery of this field, its existence was regarded as something of a mystery. For a planet to generate and sustain an overall magnetic field, it needs to have a liquid interior within which electrical currents can circulate. Astronomers had thought that, because Mercury is so much less massive than the Earth, its iron core would have cooled down and solidified by now. However, recent evidence indicates that it may be at least partially molten, in which case the presence of a magnetic field is easier to understand.

Venus

With a diameter of 12,104 km, Venus is almost the Earth's twin in terms of size but, curiously, it rotates round its axis in a retrograde direction (in the opposite direction to the Earth's rotation and to the direction in which the planets move round the Sun) in a period of 243 days.

The planet is totally and permanently covered by cloud layers that reflect about 76% of the incoming sunlight and this, together with its relative proximity to the Earth, ensures that Venus, when at its most brilliant, is by far the brightest of the naked eye planets. The venusian clouds consist predominantly of droplets of sulphuric acid, and the atmosphere, which consists mainly of carbon dioxide (96.5% by mass) and nitrogen (about 3.5%) is so heavy that it exerts a pressure at ground level that is more than 90 times greater than the atmospheric pressure at the surface of the Earth. This thick atmospheric blanket creates an extreme greenhouse effect, which has raised the temperature at the planet's surface to 735 K (462 °C) (Fig. 5.5).

Radar measurements, which penetrate the planet's thick cloud layers, have revealed a fascinating landscape, with broad, gently undulating plains, craters, valleys, chasms, and several elevated regions reminiscent of continents. The highest features are the Maxwell Mountains (Maxwell Montes), which rise to some 11 kilometres above the mean surface level. There are numerous gently sloping 'shield' volcanoes, which appear to be extinct, although it is possible that some slight activity continues. The undulating lowlands, which appear to have been flooded by lava, contain sinuous channels, some of which are longer than the longest rivers on

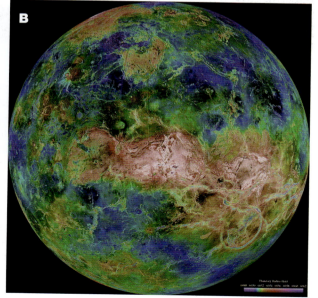

Figure 5.5 Venus: **A**, this visible/ultraviolet image shows stripe-like structures in the clouds that envelop the entire planet. Image credit: ESA/VIRTIS/INAF-IASF/Obs. De Paris-LESIA. **B**, radar map of the surface, colour coded to show differences in elevation. The light-coloured region extending across the centre is Aphrodite Terra, an elevated plateau which measures about 10,000 km by 3000 km. Image credit: NASA/JPL/USGS.

the Earth. Impact craters are present in considerable numbers, but they are not nearly as abundant as on the Moon or Mercury. There is no clear evidence of plate tectonics, so in that respect, the structure and dynamics of the planet differ markedly from the Earth.

Mars

At its mean distance from the Sun of 1.52 AU, Mars has an orbital period of 687 days (1.88 years). It rotates around its axis in 24 h 37 m, and so the length of its 'day' is closely similar to our own. Furthermore, because the tilt of its axis (23° 59′) is almost identical to that of the Earth, Mars experiences a similar pattern of seasons to our own, although each season lasts nearly twice as long as its terrestrial equivalent.

With about half of the Earth's diameter and a tenth of the Earth's mass, Mars is the second smallest and the least dense of the terrestrial planets. Much of its reddish surface is littered with craters, though it is less densely peppered than the Moon or Mercury. There are several large impact basins, the most notable of which – the Hellas basin – has an overall diameter of about 2300 km. Mars shows no sign of large-scale ongoing plate tectonics, but past internal activity has pushed up a huge bulge in the crust – the Tharsis bulge – which is dominated by four huge extinct shield volcanoes. The largest of these is Olympus Mons, which stands about 24 km high. A striking feature of the martian surface is an immense canyon system – Valles Marineris – which is up to 200 km wide, 7 km deep, and about 4000 km long (Fig. 5.6).

Although no bodies of liquid water exist on the martian surface today, there is plenty of evidence to suggest that water must have flowed there at some time, or times, in the past. Such evidence includes channels that meander across the surface like dried-up river beds, layered sediments, teardrop-shaped features that appear to have been sculpted by flowing liquid, and large numbers of outflow channels that may have been formed by flash flooding. Mars may well have had a thicker atmosphere, and a warmer, wetter climate, billions of years ago when the volcanoes were active. The present-day atmosphere is tenuous, with a pressure at ground level of about 6 mb (about 0.6% of the pressure at the Earth's surface), and is composed mainly of carbon dioxide (95.3%), nitrogen (2.7%) and argon (1.6%). Because the atmospheric blanket is so tenuous, it does little to retain heat, and so the average surface temperature is only 218 K (−55°C); it can rise briefly to around 300 K (27°C) when the Sun is vertically overhead, but drops to below 200 K (−73 °C) at night and, at the winter pole, can fall as low as 160 K (−113°C).

In the vicinity of the winter pole, carbon dioxide freezes out of the atmosphere to form a layer of frost, snow and ice on top of the underlying permanent water ice cap. Consequently, the observed polar cap expands in autumn and winter and shrinks in spring and summer. Although average wind speeds at the martian surface are low, strong winds blow from time to time, and these can whip up dust storms which, on occasion, may blanket the entire planet for weeks on end.

Whether or not elementary bacterial life-forms have existed on Mars in the past, or may still exist in some locations even now, is a matter of continuing debate. One of the key goals of the ongoing exploration of the planet is to look for evidence of past or present

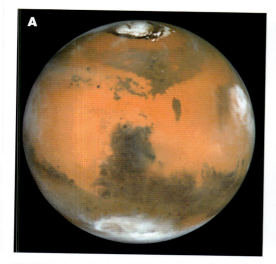

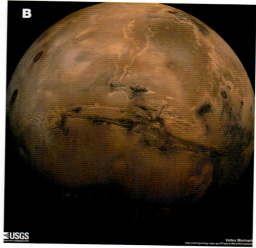

Figure 5.6 Mars: **A**, centred on the prominent dark feature, Syrtis Major, this image also displays the north polar cap (top) and (bottom) a cloud-filled basin (Hellas). Image credit: Steve Lee (University of Colorado), Jim Bell (Cornell University), Mike Wolff (Space Science Institute), and NASA/ESA. **B**, this mosaic of one martian hemisphere shows the 4,000 km long canyon system, Valles Marineris, and three giant extinct volcanoes (dark red spots, left). Image credit: USGS Astrogeology (astrogeology.usgs.gov). **C**, 'Point Lake' outcrop in Gale crater, imaged by the Curiosity Rover; the outcrop is about 2 metres wide and 0.5 metre high. Image credit: NASA/JPL-Caltech/MSSS

water and for the presence of any kind of biological material in the rocks and soil. There certainly appears to be some water locked up in the soil; for example, samples investigated by the Curiosity rover (*see* Fig. 11.12) in 2013 turned out to contain about two percent by weight of water.

Mars has two tiny natural satellites – Phobos ('Fear') and Deimos ('Dread'), both of which are irregularly shaped bodies with maximum diameters of 27 km and 15 km, respectively.

Jupiter

With a mass 318 times greater than that of the Earth, Jupiter is about two and a half times as massive as all the other planets put together. With more than 11 times the Earth's diameter, its huge globe could contain well over a thousand bodies the size of our planet, but its mean density is only about a quarter of the Earth's and is similar to that of the Sun; this implies that its composition is very different from that of the terrestrial planets. Despite its huge size, Jupiter rotates much more rapidly than the Earth, in a period of about 9 h 50 m. Because of its rapid rotation, Jupiter bulges out at its equator and is flattened at the poles, its equatorial diameter of 142,800 km being about six percent greater than its polar

diameter. This degree of flattening indicates that the interior of the planet is fluid.

Like the Sun, Jupiter is composed primarily of hydrogen and helium. It is believed to have a hydrogen-rich atmosphere that extends down to a depth of about 1000 km below the visible cloud tops (Fig. 5.7). Beneath this is a layer of liquid hydrogen some 16,000 km deep. Deeper down, where the temperature rises to more than 10,000 K and the pressure exceeds 4 million Earth atmospheres, hydrogen behaves like a liquid metal, and therefore becomes a good electrical conductor. The **liquid metallic hydrogen** zone is at least 40,000 km deep. At the centre of the planet, there is likely to be a rocky-metallic core, possibly coated with ice, with an estimated mass in the region of 15 Earth-masses. Jupiter has a powerful magnetic field, ten times stronger at its cloud tops than the magnetic field at the surface of the Earth,

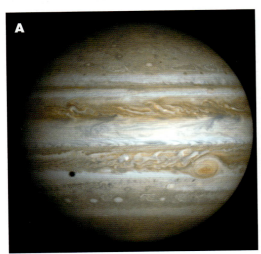

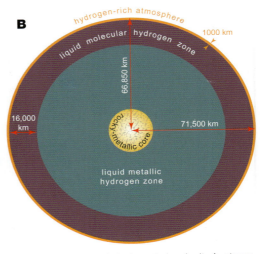

Figure 5.7 Jupiter: **A**, this image shows the dark cloud belts and lighter zones that characterize Jupiter's atmosphere, together with (lower right), the Great Red Spot and (lower left) the shadow cast by one of the Galilean satellites. Image: NASA/JPL/University of Arizona. **B**, this cross-section shows the principal regions in the interior of the gas giant, Jupiter. The size of the rocky-metallic core is very uncertain.

which is generated and sustained by circulating currents in its rapidly rotating liquid metallic hydrogen zone.

The planet's visible disc is dominated by bands of cloud – darker belts alternating with lighter zones – that wrap all the way round the planet and run parallel to its equator. Although numerous spots and storms occur in the cloud belts, the only really long-lived feature is the Great Red Spot, a rotating weather system that measures some 25,000 km long by about 12,000 km wide.

Jupiter is surrounded by an exceedingly faint system of rings and by a retinue of at least 67 natural satellites, most of which are less than 10 km in diameter. However, its four major moons – the **Galilean satellites** (which were first seen by Italian astronomer Galileo in the winter of 1609–10) – are comparable with, or larger than, the Earth's Moon. In order of distance from the planet they are: Io, Europa, Ganymede and Callisto. Ganymede is significantly larger than the planet Mercury. Europa, the smallest of the four, has a smooth icy surface, which is covered by numerous shallow cracks. This surface appears to float on top of an ocean of slushy water, within which conditions may be suitable for the existence of some kind of elementary life. Io's surface is completely devoid of impact craters and has been shaped by, and is continually being modified by, ongoing volcanic eruptions and lava flows.

Saturn

With a diameter nine times larger, and a mass 95 times greater, than the Earth, Saturn, too, is a giant of the Solar System, but with a mean density of just 700 kg/m³ (0.7 times the density of water), is by far the least dense of the planets. Although it has a slightly longer rotation period than Jupiter (10 h 14 m at the equatorial cloud tops and 10 h 39 m internally) it is more markedly flattened than is Jupiter, its equatorial diameter being about ten percent greater than its polar diameter. Saturn's chemical composition and internal structure are believed to be similar to that of Jupiter, and, like Jupiter, it has a powerful magnetic field and an extensive magnetosphere. Saturn's cloud belts and weather systems appear more muted than those of Jupiter, but major storms erupt from time to time – one of the most notable recent ones, which appeared in its northern hemisphere in 2011, eventually stretched nearly all the way round the planet (Fig. 5.8).

Its most distinctive feature is its magnificent system of rings. The main part of the ring system consists of three principal components: ring A, with an outer diameter of about 273,500 km, is separated from ring B – the brightest and widest ring – by a gap called the Cassini division, which is about 5000 km wide. Ring C, the innermost of the readily visible rings, is faint, tenuous and dusky. A fainter dusty ring (ring D) lies closer to the planet, and several faint rings lie beyond the A-ring, the outermost of which (the Phoebe ring) has a radius of about 13,000,000 km. The rings are composed of billions of particles of ice, dust and ice-coated rock, all orbiting independently above the planet's equator. Because the rings lie in the plane of Saturn's equator, which is tilted to the plane of its orbit by an angle of 27°, the aspect of the rings as seen from the Earth changes as the planet moves round the Sun. Every 13–15 years, the rings appear edge-on (this happened most recently in 2009). Each face is

A

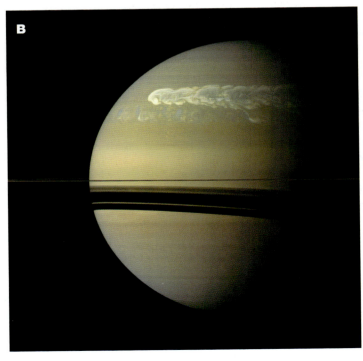

B

Figure 5.8 Saturn: **A**, this view shows the cloud belts, the ring system (including the prominent gap called the Cassini Division) and the shadow of the planet on the rings (lower left). Image credit: NASA, ESA and Erich Karkoschka (University of Arizona). **B**, The rings seen edge-on, with their shadow below, and a huge elongated storm in the northern hemisphere as seen in February, 2011. Image credit: NASA/JPL–Caltech/Space Science Institute.

displayed to best effect, alternately, midway between edge-on presentations.

Sixty-two satellites have been positively identified. Most of them are small, with diameters of 1–500 km, four are of moderate size (1000–1500 km in diameter) and one, Titan (diameter 5150 km), is larger than the planet Mercury, and is unique among planetary satellites in having a thick atmosphere, composed mainly of nitrogen (98%) and methane (about 1.4%), with a pressure at ground level equivalent to 1.5 Earth atmospheres. At its surface, where the temperature is about 95 K (−178 °C), methane can exist as a solid, liquid or gas. Radar measurements made by the Cassini spacecraft have revealed mountains, valleys, and lakes of liquid hydrocarbons such as ethane and methane on a surface that appears to be composed mainly of methane ice. Of the other satellites, Enceladus is particularly interesting; plumes of water vapour and salty ice spew out from its surface, and it seems likely that large quantities of liquid water (and perhaps conditions suitable for some kind of life) exist beneath its thin, icy crust.

Uranus and Neptune

All of the planets out to Saturn have been known, as naked eye objects, since the dawn of recorded history. The next two were discovered in the telescopic era – Uranus in 1781 and Neptune in 1846.

Orbiting at a distance of 19.2 AU, in a period of 84 years, Uranus has just under four times the Earth's diameter and 14.5 times the Earth's mass. A distinctive feature of Uranus is its extreme axial inclination (98°), which implies that its rotational axis lies almost in the plane of its orbit. This leads to a curious pattern of seasons whereby each pole experiences, alternately, 42 years of continuous sunlight and 42 years of darkness.

Uranus has a deep atmosphere composed primarily of hydrogen (about 83% by volume) helium (about 15%) and methane (about 2%). However, hydrogen makes up only about 15% of the overall composition of the planet; about 60% of its mass is believed to consist of ices (substances that would be expected to exist in solid form in the cold conditions of the outer Solar System) – water, methane and ammonia. Its interior may have three distinct layers: an iron-silicate core, comparable in size to the Earth, surrounded by an ocean of liquid or slushy ices, overlain by a deep hydrogen-rich atmosphere. Its atmosphere has a rather bland appearance because overlying hazes obscure its cloud layers (Fig. 5.9).

Uranus is surrounded by a system of 13 faint, narrow rings, most of which are just a few kilometres wide. It has 27 known moons, five of which are medium-sized bodies with diameters in the range 470 km to 1580 km; the others are small, with diameters ranging from 18 to 162 km

Neptune, the most distant planet, is marginally smaller than Uranus, but is the more massive of the two (17.2 Earth masses). Its deep atmosphere is similar in nature and composition to that of Uranus, with the planet having a generally bluish hue. It contains parallel bands of cloud, broadly similar to those seen on Jupiter and Saturn, but with far less contrast and apparent structure. Neptune's interior is believed to be broadly similar to that of Uranus.

The planet is surrounded by five faint rings, and has 14 satellites, most of which are very small. Triton, the largest, has a diameter of 2705 km and, with a surface temperature

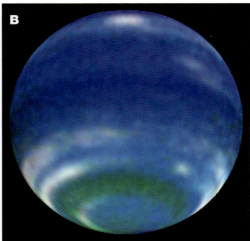

Figure 5.9 A, Uranus, showing its muted cloud belts, four of its faint rings and six of its moons. Image credit: NASA/ESA and Erich Karkoschka, University of Arizona. **B**, Neptune, showing a predominance of bright cloud features in the southern hemisphere. Image credit: NASA/ESA, L. Sromovsky, and P. Fry (University of Wisconsin-Madison).

of 37 K (-236 °C), is the coldest Solar System body that has been investigated so far by spacecraft. It has an exceedingly tenuous nitrogen atmosphere and its polar cap, which consists of frozen nitrogen, is streaked by hydrocarbon-rich material ejected from geysers that erupt through the icy crust. Unique among major planetary satellites, Triton revolves around Neptune in a retrograde direction (opposite to the planet's rotation) and is spiralling slowly inwards.

Dwarf planets

In 2006, the International Astronomical Union defined a new category of object – the **dwarf planet**. In essence, a dwarf planet is a body that travels independently around the Sun, that is sufficiently massive for its own gravity to have pulled it into a near-spherical shape, but is not massive enough to have swept all other bodies out of the region of space through which it is travelling. A consequence of accepting this definition is that Pluto – a body that, with a diameter of 2306 km, is considerably smaller than the Moon, but that previously had been regarded as the ninth planet – had to be downgraded to the status of dwarf planet. Five dwarf planets have been officially recognized so far (and there are several more candidates), one of which – Eris – is about 25% more massive than Pluto and nearly twice as distant. Estimates suggest that at least several hundred dwarf planets may lie beyond the orbit of Neptune.

Table 5.1 Planetary data

planet	semimajor axis of orbit		sidereal orbital period	synodic period (days)	equatorial diameter (km)	mass (kg)	axial rotation period
	(km)	AU					
Mercury	5.791×10^7	0.387	87.97 d	115.88	4,879	3.3×10^{23}	58.65 d
Venus	1.082×10^8	0.723	224.70 d	583.91	12,104	4.87×10^{24}	243.0 d
Earth	1.496×10^8	1.000	365.26 d	n/a	12,756	5.98×10^{24}	23.93 h
Mars	2.279×10^8	1.524	686.98 d	779.95	6,787	6.42×10^{23}	24.62 h
Jupiter	7.784×10^8	5.203	11.86 yr	398.87	142,800	1.90×10^{27}	9.8 h
Saturn	1.427×10^9	9.537	29.46 yr	378.06	120,660	5.69×10^{26}	10.2 h
Uranus	2.871×10^9	19.19	84.01 yr	369.64	51,118	8.86×10^{25}	17.9 h
Neptune	4.498×10^9	30.07	164.8 yr	367.48	49,528	1.02×10^{26}	19.1 h

This table is based primarily on data from NASA's Planetary Data System. Note that the axial rotation of Venus is retrograde.

Asteroids

The **asteroids**, or **minor planets**, are small bodies, ranging in diameter from 950 km down to less than 1 km, which revolve round the Sun in their individual orbits. The largest one, Ceres, orbits at a distance of 2.77 AU in a period of 4.6 years. With a diameter of 950 km and a near-spherical shape, it has recently been re-classified as a dwarf planet. Only seven of the asteroids have diameters greater than 300 km; two hundred or so are larger than 100 km, and it is estimated that there are several hundred thousand with diameters in excess of 1 km. Most asteroids, apart from a few of the largest, are irregular in shape (Fig. 5.10).

Most of the known asteroids lie within the main asteroid belt, which extends from about 2.0 AU to about 3.3 AU from the Sun. In addition, some asteroids orbit well outside the main belt, some inhabit the space between the orbits of Saturn and Neptune, and others are located beyond the orbit of Neptune. The

Figure 5.10 This view of asteroid Vesta (maximum diameter, 580 km), by the Dawn spacecraft, shows craters and, at the bottom, a mountain more than twice the height of Mt. Everest. Image: NASA/JPL-Caltech/UCAL/MPS/DLR/IDA.

trans-Neptunian bodies form an extended disc-shaped population of ice-rich asteroids – known as the **Kuiper belt** – that stretches from around 30 AU to about 100 AU from the Sun. Some asteroids come close to, or cross, the orbit of the Earth. Collectively, these objects are known as **near-Earth objects** (**NEOs**), and current estimates suggest that nearly a thousand of them have diameters greater than 1 km. By convention, an asteroid (or a comet) is considered to be an NEO if its perihelion distance is less than 1.3 AU. An asteroid that follows an orbit which can bring it to within 0.05 AU of the Earth is called a **potentially-hazardous asteroid** (**PHA**).

The asteroids are debris left over from the formation of the Solar System some 4.6 billion years ago, and represent material that never aggregated together to form a single planetary body.

Comets

A major **comet** can be a dramatic sight. With its fuzzy head and long tail, or tails, it appears to hang like a sword in the heavens, changing its position relative to the stars night by night before vanishing as mysteriously as it arrived. Little wonder that comets were regarded with a mixture of awe and fear, and seen as portents of disaster in ancient and mediaeval times.

As English astronomer, Edmond Halley (1656–1742), discovered, comets are bodies that travel round the Sun on elongated orbits under the action of the Sun's gravity. They are divided into two very broad categories – short-period (with orbital periods of less than 200 years) and long-period (with orbital periods greater than 200 years). Many of the long-period comets have periods of thousands, or even millions, of years, and their returns cannot be predicted with any confidence. Comets are usually named after their discoverer(s), and periodic comets – which

have well determined orbits and which return to perihelion at regular intervals – are given the prefix 'P/'; for example, Halley's comet is denoted by 'P/Halley'.

The main features of a well-developed comet are the **nucleus**, **coma** and **tail** (Fig. 5.11). The nucleus is a lump of ice, dust and rocky material (often referred to as a 'dirty snowball') that is typically a few kilometres in diameter. Each time the nucleus approaches perihelion, gas and dust evaporate from its surface to form a cloud, called the coma, which can grow to a diameter of 100,000 km or more. A comet will often display two types of tail, a gas (or more properly, ion) tail and a dust tail (Fig. 5.12). Ultraviolet light from the Sun removes electrons from atoms and molecules in the coma, so creating electrically charged ions that are swept away by the fast-moving solar wind to create the ion tail (a Type I tail), which points almost directly away from the Sun. Sunlight also exerts a tiny, but finite, pressure (**radiation pressure**) that pushes dust particles out of the head of the comet at somewhat lower speeds. Consequently, the dust particles lag behind the head of the comet, creating a curved dust tail (a Type II tail). Both types of tail follow the head of the comet on the way in towards perihelion and precede it on the way out. Fully developed cometary tails can extend for millions, or even hundreds of millions of kilometres.

Because a comet loses material each time it makes a close approach to the Sun, the active lifetime of a short-period comet may be no more than a few tens of thousands of years; if their numbers were not being continually replenished, the stock of short-period comets would long since have

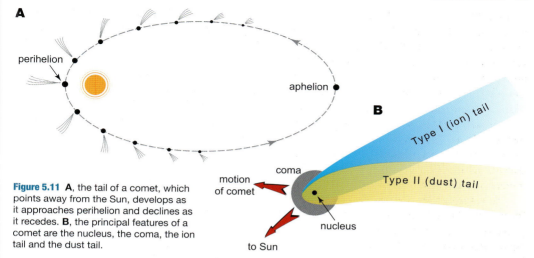

Figure 5.11 A, the tail of a comet, which points away from the Sun, develops as it approaches perihelion and declines as it recedes. **B**, the principal features of a comet are the nucleus, the coma, the ion tail and the dust tail.

been exhausted. Many of the short-period comets are believed to come from the Kuiper belt, whereas long-period comets, and comets that are making their first close encounter with the Sun, originate from a much larger distribution of icy objects – called the **Oort cloud** – which extends out to about 50,000 AU (a light year) from the Sun.

Meteors and meteoroids

Interplanetary space contains vast numbers of particles, called **meteoroids**, which range in size from microscopic grains to millimetre-, centimetre- and metre-sized objects. When meteoroids of millimetre-size or larger plunge into the Earth's atmosphere they are heated to incandescence by friction and are vaporized, creating a short-lived trail of light (which seldom lasts for much more than a second) that is called a **meteor** (or, mis-leadingly, a 'shooting star').

Meteors may be divided into two major classes – sporadic and shower. **Sporadic meteors**, which can appear at any time from any direction, are caused by particles approaching the Earth from random directions. A **meteor shower**, on the other hand, occurs when the Earth passes through a stream of meteoroids, all of which are moving along the same orbit, or closely similar orbits. The meteoroids that populate these streams are particles that have been driven from the heads of comets. Because these meteoroids are following parallel paths, perspective causes them all to seem to diverge from a common point on the sky, called the **radiant** (Fig. 5.13). A shower is usually named after the constellation within which its radiant lies. A classic example is the Perseid shower, which reaches a peak around 12 August each year and which appears to emanate from a point in the constellation Perseus.

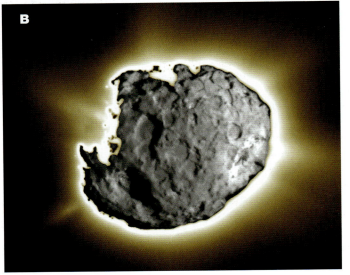

Figure 5.12 **A**, Comet Hale-Bopp, on 14 March 1997, showing the dust tail fanning out to the right, and the blue ion tail pointing straight away from the Sun. Image credit: ESO/E. Slawik. **B**, the 5-km wide nucleus of comet Wild 2, imaged by the Stardust spacecraft, showing its cratered surface and emerging jets of gas and dust. Image credit: NASA/JPL-Caltech.

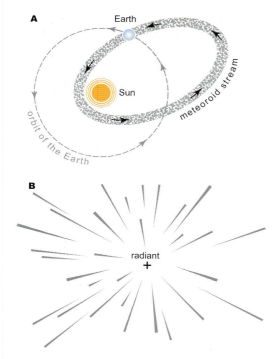

A

Earth

Sun

orbit of the Earth

meteoroid stream

B

radiant
+

Figure 5.13 Meteor shower: **A**, when the Earth passes through a meteoroid stream, meteors appear to spread out (**B**) from the radiant.

Meteorites

Meteorites are more substantial bodies, which survive (at least in part, or as fragments) their fiery passage through the atmosphere and can reach the ground. As a meteorite plunges through the atmosphere, it is heated to incandescence, and melted material streams away from its surface. Its passage through the atmosphere is marked by a brilliant ball of light and streaming tail, known as a **fireball**, which may on occasion be brighter than the full Moon, and very occasionally may briefly rival the Sun.

Meteorites are divided into three principal types: stony, stony-iron, and iron. Stony meteorites are composed of rock; stony-irons consist of roughly equal quantities of rock and iron-nickel; and irons consist almost entirely of a mixture of iron and nickel (predominantly iron). Stony meteorites, which are more fragile, tend to break into fragments as they plunge to Earth, but irons are much more likely to land intact. The largest single meteorite that has been found on the surface of the Earth weighs about 60 tonnes and lies, where it fell, in Namibia. The compositions of the different types of meteorites seem to mirror those of the different types of asteroids. It seems virtually certain that the great majority of meteorites originate from the asteroid belt, and that all but a few are fragments chipped off asteroids as a result of collisions between them.

Meteorites provide us with samples of asteroidal material. By studying the relative proportion of radioactive elements and their decay products in individual crystals embedded within meteorites, it is possible to calculate how much time has elapsed since those bodies solidified. For the great majority of meteorites, this turns out to be around 4.55 billion years; meteorites give us our best guide to the age of the Solar System, and furnish us with samples of the oldest solid material in our planetary system.

Impacts and hazards

Asteroids, cometary nuclei, and exceptionally massive meteorites can strike the ground with sufficient energy to excavate craters. The best-known terrestrial impact structure is the Barringer crater (commonly called 'Meteor Crater') in Arizona. Measuring 1360 metres across and with a depth of 160 metres, it is believed to have been created some 50,000 years ago by the impact of a 70,000 tonne iron meteorite. But this is a very modest crater compared to some that have been identified – often highly eroded or buried – at various locations on planet Earth. For example, the remnants of a 180-km crater in the Yucatan Peninsula of Mexico indicate that a 10-km asteroid struck that part of the Earth around about 65 million years ago. Impacts on this scale, or larger, have produced profound changes on the Earth's climate and the evolution of life. One popular hypothesis is that the Yucatan impact triggered the chain of events that killed off the dinosaurs. On a positive note, though, it may well be that most of the Earth's inventory of water was delivered by impacting comets very soon after the planet was formed.

Although estimates of the frequency of massive impacts vary considerably, it seems likely that the Earth will be struck by a 10-km sized object approximately once every hundred million years, a 1-km object perhaps once every 200,000 years and a 10–100-metre object once a century. The Earth has experienced several near misses in recent years, with bodies with diameters of tens of metres occasionally passing closer than the Moon. A small asteroid or cometary nucleus that exploded in the atmosphere rather than hitting the ground directly is believed to have been responsible for a blast that flattened thousands of square kilometres of forest in the Tunguska region of Siberia in 1908. Much more recently, in 2013, a similar – though more modest – event occurred when an incoming object, estimated to weigh about 10,000 tonnes, exploded in a dazzling fireball close to the Russian town of Chelyabinsk, shattering office windows and causing many hundreds of injuries (mainly from flying glass).

Our growing awareness of the Earth's vulnerability to massive impacts has spawned a number of surveys to track down, and plot the orbits of, NEOs that have the potential to collide with, and damage, our planet. Strategies for deflecting or destroying hazardous asteroids and cometary nuclei are being actively investigated (not just in movies!), for the Earth is, after all, a target in a cosmic shooting gallery.

6   Stars – their properties and variety

Stars are incandescent globes of gas, which are broadly similar in nature to our Sun. Some are larger, some smaller, some hotter, some cooler, some are far more luminous, while others are, compared to the Sun, mere celestial glow-worms.

Brightness and magnitudes

The brightness of a star as seen in the sky is described by a quantity called **apparent magnitude** – a measure of the amount of light arriving at the Earth that owes its origins to the work of the second century Greek astronomer, Hipparchos, who divided the naked-eye stars into six classes, or magnitudes, with the brightest being designated first magnitude and the faintest, sixth magnitude. The modern version of the **magnitude scale** is a logarithmic one, defined in such a way that a difference of five magnitudes (e.g. the difference between magnitude 1 and 6) corresponds to a factor of 100 in brightness; for example, a star of magnitude 6 is 100 times fainter than a star of magnitude 1. A step of one magnitude corresponds to a brightness difference equal to the fifth root of 100, which is 2.512. Thus, compared to a star of magnitude 1, a star of magnitude 2 is 2.512 times fainter, a star of magnitude 3 is $(2.512)^2 = 6.31$ times fainter, and a star of magnitude 6 is $(2.512)^5 = 100$ times fainter.

Stars fainter than naked-eye visibility have magnitudes greater than 6. For example, a star of magnitude 11 is 100 times fainter than a star of magnitude 6 and 10,000 times (100×100) times fainter than a star of magnitude 1. Celestial objects that are brighter than first magnitude have fractional, zero or negative magnitudes. Hence a star of magnitude 0 is 2.512 times brighter than a star of magnitude 1, a star of magnitude 1 is 6.31 times brighter, and so on. Sirius – the brightest star in the sky – has an apparent magnitude of −1.46. The planet Venus, at its greatest brilliancy, reaches magnitude −4.4 (more than a hundred times brighter than a star of magnitude 1); the full Moon attains magnitude −12.6, and the Sun's apparent magnitude is −26.7. Seen in our skies, the Sun is about 33 magnitudes (about 16 trillion times) brighter than the faintest naked-eye star. At the other end of the scale, the faintest objects detectable by the most powerful telescopes are of magnitude +29, which is more than a billion times fainter than the naked-eye limit.

Distance and parallax

The fundamental method of measuring the distances of stars is trigonometrical **parallax**. In principle, if we were to measure the precise position of a nearby star in January, say, when the Earth is on one side of the Sun, and again six months later (in July), when the Earth is on the opposite side of the Sun, then because these observations have been made from locations separated by the diameter of the Earth's orbit (300,000,000 km), we should be able to discern a slight difference in its position on the sky. An everyday analogy is to hold out one finger at arm's length, close your left eye and then, using your right eye, line up your finger with a distant object. Then, without moving your finger, close the right eye and open the left. Your finger will no longer be lined up with that object because your eyes, being separated by a few centimetres, are looking in slightly different directions towards your nearby finger.

The maximum angular displacement of a star from its mean position on the sky is called its **annual parallax** (Fig. 6.1). If we can measure this tiny angle then, knowing the distance between the Earth and the Sun, we can use trigonometry to work out the distance between the Sun and the star. No star is close enough for its annual parallax to be as great as 1 **arcsec** (an arcsec is a second of angular measurement, which is 1/3600 of one degree). A unit of distance, which relates directly to the measured parallax of a star, is the **parsec**, which is the distance at which a star would have an annual parallax of 1 arcsec, and is equal to 206,265 AU (30.9 trillion km) or 3.26 light years. The distance of a star expressed in parsecs is equal to the reciprocal of its parallax (in arcsec). For example, Proxima Centauri, with an annual parallax of 0.772 arcsec, is at a distance of $1/0.772 = 1.30$ parsecs, which is equivalent to 4.2 light years.

Because it is hard to make precise positional measurements while looking through the Earth's turbulent atmosphere (*see* Chapter 11), the smallest parallaxes that can be measured from ground level by conventional means, and with reasonable accuracy, are around 0.03 arcsec, which corresponds to distances of about 30 parsecs (approximately 100 light years). To get to greater distances, astronomers

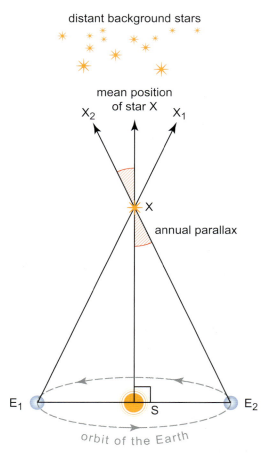

Figure 6.1 As the Earth moves round its orbit from E1 to E2, the observed position of star X shifts from X1 to X2. The annual parallax of the star is its maximum displacement from its mean position (angle SXE2).

have to use more indirect methods, which rely on being able to deduce the true **luminosity** (light output) of a star from some other measurable property such as its spectrum. Then, because the brightness of a star diminishes with the square of its distance (if the distance is doubled, the apparent brightness is reduced to a quarter), its distance can be calculated by comparing its observed apparent brightness with its assumed luminosity.

But, so far as parallax measurements are concerned, space-borne instruments can do much better than ground-based ones. The Hipparcos satellite, launched in 1989, determined the parallaxes of 120,000 stars with a measurement accuracy of about 0.001 arcsec, yielding reasonably good values of distance out to a thousand light-years or so. The Gaia spacecraft, which was launched in December 2013, has been designed to improve on that performance by a factor of a hundred. One

of the aims of the mission is to measure the parallaxes of about a billion stars with accuracies in the region of 20 *millionths* of an arcsec; this will enable astronomers, for the first time, to use parallax to determine the distances of stars out to distances in the region of 10,000 to 100,000 light years (Fig. 6.2).

Luminosity and absolute magnitude

Apparent magnitude is a measure of the amount of light arriving at the Earth from a star but, on its own, is no guide to that star's actual light output. A star of modest luminosity will appear bright if it is fairly close to us, whereas

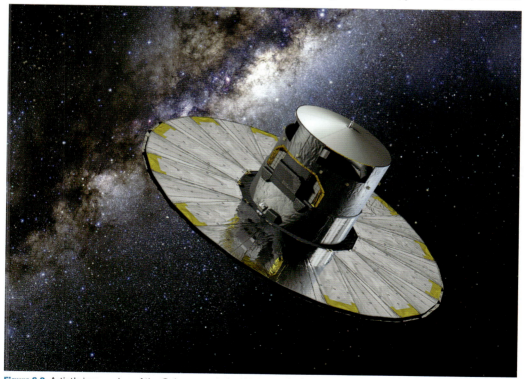

Figure 6.2 Artist's impression of the Gaia spacecraft, which aims to measure the positions, motions and properties of a billion stars. Image credit: ESA-D. Ducros, 2013.

a star that is inherently highly luminous will appear faint if it lies at a very great distance. In order to compare the inherent luminosities of stars on the magnitude scale, astronomers use a quantity called **absolute magnitude**, which is defined as the apparent magnitude that a star would have if it were placed at a standard distance of 10 parsecs from the Earth. The absolute magnitude of the Sun (apparent magnitude –26.7) is +4.8, so if the Sun were removed to a distance of 10 parsecs (32.6 light years), it would be a rather dim star, not a lot brighter than the naked eye limit. By way of comparison, Deneb, the brightest star in Cygnus, which has an apparent magnitude of 1.25, has an absolute magnitude of –7.5; this reflects the fact that it is about 80,000 times more luminous than the Sun. By way of contrast, Proxima Centauri (the nearest star) has an absolute magnitude of +15.5, which implies that its inherent luminosity is a little over one twenty-thousandth of the Sun's.

More broadly, we can define **luminosity** to be the total amount of energy that a star radiates per second from its entire surface. In the case of the Sun, this is 3.86×10^{26} watts. The most luminous stars radiate more than a million times as much power as the Sun, whereas the least luminous have only a few hundred-thousandths of the Sun's output. Luminosity depends on the surface temperature and the total surface area of the star. If two stars have the same surface temperature, the larger of the two has the greater surface area, and therefore will have the greater luminosity. Conversely, if two stars have the same radius and surface area, but one is hotter than the other, the hotter of the two will be the more luminous.

Colour, temperature and size

The colour of a star is an indicator of its surface temperature. Cool stars – with temperatures in the region of 3000 K (for example, Betelgeuse in Orion) – appear red, whereas hotter stars such as the Sun (approximately 6000 K) look yellowish. Stars with surface temperatures in the region of 10,000–12,000 K (such as Rigel, in Orion) appear white, while the hottest stars – with temperatures of 30,000 K or more – appear bluish. The wavelength at which a hot body, such as a star, radiates most strongly is inversely proportional to its surface temperature. For example, a star with a temperature of 5800 K will emit most strongly at a wavelength of 500 nanometres, which is in the middle of the visible spectrum. If its surface temperature is 2900 K, it will shine most strongly at a wavelength of 1000 nm, which is in the near infrared part of the spectrum; but to our eyes, it will appear red. Or again, if its surface temperature were 11,600 K, it would radiate most strongly at 250 nm, in the ultraviolet; such a star would look bluish-white.

Because stars are so far away, they look like points of light even when viewed through very large telescopes. Only in a very few cases can the star be seen as a tiny disc, the diameter of which can be measured directly (assuming its distance is well known). However, the radius of a star can be calculated if its true luminosity and temperature are known. The largest stars of all (**red supergiants**) are bigger than the orbit of Jupiter, whereas shrunken stars called **white dwarfs** are only about the size of planet Earth, and extremely compressed stars, called **neutron stars** (*see* Chapter 7), are very much smaller again.

Interpreting the spectrum

The rainbow band of colours which is seen when white light – a mixture of all the different colours – is spread out into its constituent wavelengths by a **spectroscope** (*see* Chapter 11) is called a **continuous spectrum**. The **spectrum** of a typical star, consists of a continuous spectrum (a rainbow band of colours), upon which is superimposed a pattern of dark lines (**absorption lines**), each line occurring at its own distinctive wavelength. In contrast, a glowing cloud of low-density gas emits light at certain specific wavelengths only, and its spectrum consists of a set of bright lines (**emission lines**) (Fig. 6.3).

The origin of these lines is related to the structure of the atom. A typical atom consists of a central **nucleus** – composed of heavy particles with positive electrical charge called **protons**, and particles with similar mass but zero electrical charge called **neutrons** – which is surrounded by a number of lightweight, negatively charged particles called **electrons**, which orbit around the nucleus rather like planets around the Sun. The simplest and lightest atom is the hydrogen atom, which consists of a single proton and a single electron. According to the Bohr model of the atom, developed in 1913 by the Danish physicist Neils Bohr, the single orbiting electron

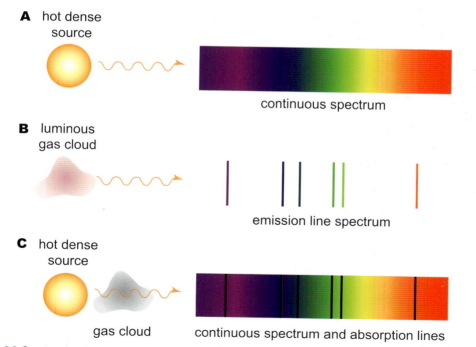

A hot dense source

continuous spectrum

B luminous gas cloud

emission line spectrum

C hot dense source

gas cloud continuous spectrum and absorption lines

Figure 6.3 Spectra: **A**, a hot, dense source of light emits a continuous spectrum. **B**, a luminous cloud of low-density gas emits at particular wavelengths only (an emission line spectrum). **C**, when a continuous spectrum passes through a gas cloud, dark absorption lines are superimposed.

can exist in one of a number of different permitted orbits, each of which corresponds to a particular energy level, with the innermost orbit having the lowest energy and the outermost, the highest. If an electron in one of the lower levels absorbs a quantity of energy equal to the difference between that level and a higher one, it will jump up to the higher one. It can achieve this by absorbing a **photon** (a 'particle' of light) that has exactly the right energy and wavelength (the higher the energy, the shorter the wavelength). Conversely, when an electron drops down from a higher level to a lower one, the atom will emit a photon whose energy is equal to the energy gap between the two levels. Each of the possible transitions (jumps) between the various permitted energy levels corresponds to the absorption or emission of different wavelengths of light.

As light from the hot, dense interior of a star passes out through its photosphere and atmosphere, the combined absorbing effect of large numbers of hydrogen atoms imprints a set of dark absorption lines in its spectrum – a set that is unique to hydrogen. In similar fashion, atoms of other chemical elements imprint their own 'fingerprint' patterns of dark lines on the spectrum of the star. By identifying and analysing the various series of lines in the spectra of stars, astronomers can determine their chemical compositions. Various other factors (for example, temperature) exert an influence on whether particular lines will be present or absent, prominent or weak, in the spectrum of a star. Consequently, detailed interpretation of a star's spectrum can provide information not only about its chemical composition but also about other properties such as its temperature, atmospheric pressure, rate of rotation, magnetic field strength, and the speed at which it is moving towards or away from us. Analysis of stellar spectra provides the key to determining the physical nature and chemical properties of stars.

Classifying stars

Stars are classified into different **spectral types** according to which lines are present or absent, prominent or 'weak', in their spectra. The main classes, in order of decreasing temperature, are labelled O, B, A, F, G, K, M. O-type stars are the hottest, with surface temperatures in excess of 30,000 K, and appear blue in colour. A-type stars, with surface temperatures of around 10,000 K, appear white; G-type stars, with temperatures in the region of 6000 K, are yellowish (the Sun is a G-type star); K-type stars, with temperatures of around 4500 K, are orange; and cool M-type stars, with surface temperatures in the region of 3000 K, look red. Further classes, which have been added to the sequence in recent years to cater for very cool stars and objects known as **brown dwarfs** (*see* Chapter 7) are L (temperatures from 2400 K down to 1300 K), T (1300 K–500 K) and Y (less than 500 K). Each class is subdivided into ten sub-classes – again in decreasing temperature order – from 0 to 9. Thus a star like the Sun, which is of type G2, is slightly cooler than a G0 star (Fig. 6.4).

The Hertzsprung-Russell diagram

Information about the temperatures and luminosities of stars can be displayed in a very useful way on a **Hertzsprung-Russell diagram** (or **H-R diagram**). Diagrams of this kind, which were devised in the early part of the twentieth century by the Danish astronomer

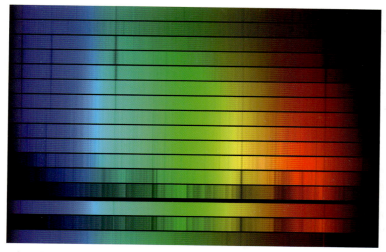

Figure 6.4 Spectral classes: this image compares the spectra of different classes of star. From top to bottom (hot to cool), the 13 spectra shown are of types: O6.5, B0, B6, A1, A5, F0, F5, G0, G5, K0, K5, M0 and M6. The three additional spectra at the bottom are special cases. Image credit: NOAO/AURA/NSF.

Ejnar Hertzsprung and the American astronomer, Henry Norris Russell, plot luminosity (or absolute magnitude) on the vertical axis and temperature (or spectral class) along the horizontal axis. Luminosity increases from bottom to top along the vertical axis and is often expressed in units of the Sun's luminosity (where the luminosity of the Sun = 1). Because, historically, the spectral classes from O to M were plotted from left to right, temperatures decrease from left to right on the H-R diagram (Fig. 6.5).

An individual star may be represented on the diagram by a point corresponding to its surface temperature (or spectral class) and its luminosity (or absolute magnitude). For example, the point representing the Sun would be placed at temperature 5780 K

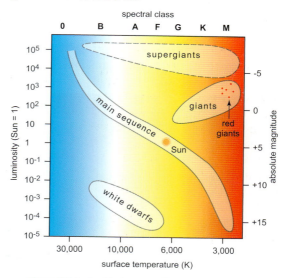

Figure 6.5 In the Hertzsprung-Russell diagram the luminosities (or absolute magnitudes) of stars are plotted against their temperatures (or spectral classes). The principal regions on the diagram are the main sequence, giants, supergiants and white dwarfs.

(or spectral type G2) and luminosity 1 (or absolute magnitude +4.8). When very large numbers of stars are plotted on the diagram, the great majority lie within a band, called the **main sequence**, which slopes down from upper left (hot, highly luminous) to lower right (cool, low luminosity). The Sun is a **main sequence star**.

Some stars lie above and to the right of the main sequence. These have higher luminosities than main sequence stars with the same surface temperature. Since stars with the same surface temperature emit the same amount of energy per second from one square metre of their surfaces, if two stars have the same temperature, but one is far more luminous that the other, the more luminous of the two must be the larger. Stars in this region of the diagram are called **giants**, prominent among which are **red giants** ('red' because they are cool, and 'giant' because they are very much larger than main sequence stars). A typical red giant has a surface temperature of around 3000 K, a luminosity 100–1000 times greater than, and a diameter 20–100 times larger than, that of the Sun (Fig. 6.6). **Supergiants** are even more luminous. For example, the red supergiant Betelgeuse is so large that if it were placed where the Sun is, all of the terrestrial planets (Mercury, Venus, the Earth and Mars) would lie within its surface; indeed, its bloated, low-density atmosphere would extend well beyond the orbit of Jupiter. Other stars lie below and to the left of the main sequence. Although some of these stars have very high surface temperatures, their luminosities are far less than those of main sequence stars with the same surface temperatures. Consequently, they must be very much smaller. With about one hundredth of the Sun's diameter, they are comparable in size to the planet Earth, and are known as **white dwarfs** ('white' because of their high temperatures, and 'dwarf' because of their small size) (Fig. 6.7).

Stellar masses range from about 150 solar masses (such stars are exceedingly rare) down

Figure 6.6 The red supergiant, Betelgeuse, in Orion, is identified by the marker (left frame). The centre frame shows a zoom in, and the right frame an exceptionally sharp image of the star (obtained by ESO's Very Large Telescope), showing a huge plume of gas extending from its surface. Image credit: ESO, P. Kervalla, Digitized Sky Survey 2 and A. Fujii.

Figure 6.7 Main-sequence star Sirius A, the brightest star in the sky, together with its faint white dwarf companion, Sirius B (lower left). Image: NASA, ESA, H. Bond (STScI), and M. Barstow (University of Leicester).

radius, packs all of its material into one millionth of the Sun's volume. Its mean density, therefore, is about a million times greater than that of the Sun.

The movements of stars

Stars move through space relative to the Sun, but their distances are so great that their changes in position are almost imperceptible. The motion of a star can be divided into two parts – **radial velocity** – the component of a star's speed directly towards, or away from, the Solar System – and **transverse velocity** – speed across the line of sight. Radial velocity can be determined by measuring the **Doppler effect** on the lines in a star's spectrum (if the star is receding, its light waves are stretched, and the pattern of dark lines in its spectrum is shifted towards the long-wave (red) end of its spectrum by an amount that is proportional to its radial velocity; this is called a **redshift**. If the star is approaching, its spectral lines are shifted towards the short-wavelength (blue) end of the spectrum – a **blueshift**).

Transverse velocity can be determined by measuring the tiny annual angular shift in a star's position, which is called **proper motion**. If the distance of the star is known, then its slow angular shift in position can be converted into a speed. The largest known proper motion is that of a dim red star called Barnard's star, which currently lies at a distance of 6.0 light years and has a proper motion of 10.3 arcsec per year. At that rate, it would take about 180 years to shift through an angle equivalent to the apparent size of the full Moon. Most stellar proper motions are very much smaller than this.

Barnard's star is moving rapidly across the line of sight, and its radial velocity is about

to about 0.08 solar masses. Most stars have masses somewhere between 10 solar masses and around one tenth of a solar mass – not a huge range; but because their diameters can differ hugely, their mean densities span an enormous range. The mean density of the Sun is about 1400 kg/m³ (1.4 times the density of water). By contrast, a red giant 50 times the size of the Sun would have sufficient volume to contain more than 100,000 bodies the size of the Sun and yet would probably have a mass quite similar to that of the Sun. Consequently, its mean density would be about one hundred thousandth of the density of the Sun (about a hundredth of the density of air at sea level on planet Earth). At the opposite end of the scale, a white dwarf, with a mass similar to that of the Sun but only one hundredth of its

140 km/s towards the Solar System. The combination of these two motions indicate that Barnard's star will make its closest approach to the Solar System in about 10,000 years' time, and will then be about 3.85 light years away from us.

Binary and multiple stars

Although there are many single stars, like the Sun, well over half of all stars are members of multiple systems containing two or more member stars. A **binary** consists of two stars that revolve around each other, bound together by their mutual gravitational attraction, in periods of time that range from less than a day to hundreds or thousands of years. The two stars travel around their common **centre of mass** – a point that lines somewhere between the two. If the two stars are of equal mass, the centre of mass will lie midway between them, whereas if one star is more massive than the other, the centre of mass will be closer to the more massive of the two; for example, if one member of the pair is twice as massive as the other, the centre of mass will be located one-third of the way from the more massive star to the less massive one. A useful analogy is to visualize the balance point of a weightlifter's bar with equal, or with unequal, weights at opposite ends (Fig. 6.8).

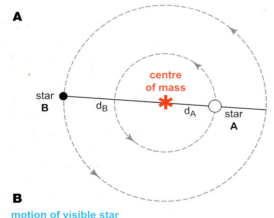

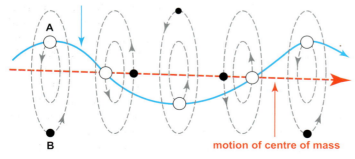

Figure 6.8 **A**, in a binary system, the more massive star (star A) is closer to the centre of mass than the less massive one (star B). **B**, with an astrometric binary, the wobbling motion of the visible star is caused by its motion round the centre of mass combined with the motion of the centre of mass through space.

Observations of binaries provide the only direct way of 'weighing' the stars. If the orbital period of, and mean distance between, the two stars can be measured, then, because the orbital period of a binary depends on the combined masses of its two component stars and the distance between them, the sum of their masses can be calculated. If their relative distances from the centre of mass can also be measured, then – because the distance of each star from the centre of mass is inversely proportional to its mass - the ratio of the masses of the two stars, and hence the mass of each one, can be obtained.

If the two stars are sufficiently far apart to be resolved (to be seen as separate points of light), the system is called a visual binary. Most binaries cannot be resolved, either because their members are too close together (or separated by too small an angle because they are so very far away from the Earth) or because one star is so much less luminous than the other that it cannot be seen. But the binary nature of what seems to be a single star can sometimes still be deduced, if it turns out to be an astrometric, spectroscopic or eclipsing binary.

In the absence of any other disturbing force, the centre of mass of a binary system travels through space in a straight line at a constant speed. If one star is visible and the other too faint to be seen, the orbital motion of the visible star around the centre of mass will cause it to wobble in a periodic fashion from side to side of this mean motion. If the slight wobble can be measured, the binary nature of the pair will be revealed: such a system is called an **astrometric binary**.

If the two stars are so close together that they appear as a single point of light, the spectrum of this 'star' will contain two sets of spectral lines; such a system is called a **spectroscopic binary** (Fig. 6.9). As the two stars revolve round their centre of mass then, at any particular instant, one star (call it A) will be approaching, while the other (B) is receding. Because of the Doppler effect, lines in the spectrum of star A will be blue-shifted by a small amount, whereas lines in the spectrum of star B will be red-shifted by a small amount. A quarter of an orbit later, both stars will be travelling at right angles to our line of sight (neither approaching nor receding), and the two sets of lines will merge into one. Thereafter, A will be receding and B approaching, so that the two sets of lines will separate out again. Likewise, if one star is too faint to make a detectable contribution to the overall spectrum, only one set of lines will be seen, but they will still shift to and fro in wavelength as the visible star revolves round the centre of mass; such a star is called a single-line binary.

If the plane of a binary's orbit is edge-on to our line of sight, then each star will pass alternately in front of its neighbour, causing eclipses. Consequently, what looks like a single star will vary in brightness in a periodic fashion. The observed brightness will be constant and equal to the combined brightness of the two stars for most of the time, but will dip down at regular intervals when one or other passes in front of its neighbour. A binary of this kind is called an **eclipsing binary** (Fig. 6.10). If the two stars are identical, each dip in brightness will be the same, whereas if one star is larger and more luminous than the other, then the **light curve** (the graph of brightness plotted against time) will show alternating dips of unequal depth (a deep one followed by a shallow one) (Fig. 6.10). The

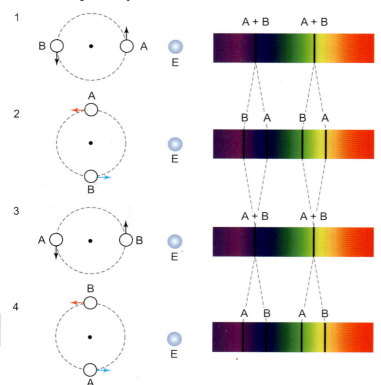

Figure 6.9 The spectrum of a spectroscopic binary contains two sets of dark lines. Viewed from the Earth (E), the two sets of lines merge when the stars are moving at right angle to the line of sight (1) and (3). When one star is approaching and the other receding (2) and (4), one set of lines is blue-shifted and the other, red-shifted.

best-known example of an eclipsing binary is the star Algol (β Persei), a second magnitude star in the constellation Perseus. During its deeper, primary eclipse, which occurs at intervals of 2.9 days, the drop in brightness is about one magnitude, which is readily apparent to the naked eye.

Many other examples of binary and multiple stars can be seen with binoculars or small telescopes, and a few can even be seen with the naked eye. Perhaps the most famous is second magnitude Mizar, the middle star of the handle of the Plough (Big Dipper), which has a fainter (fourth magnitude) companion,

Alcor, that is discernable to the naked eye under good conditions. Mizar itself is a telescopic binary, the brighter component of which (Mizar A) is itself a spectroscopic binary.

Variable stars

Some stars vary in brightness periodically, irregularly or abruptly, and these, collectively, are known as **variable stars**. Those that undergo genuine variations in luminosity are called intrinsic variables. Others, that appear to vary because of some external cause (for example, eclipsing binaries), are called extrinsic variables.

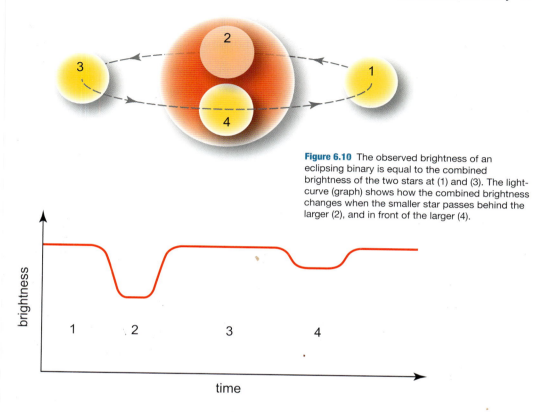

Figure 6.10 The observed brightness of an eclipsing binary is equal to the combined brightness of the two stars at (1) and (3). The light-curve (graph) shows how the combined brightness changes when the smaller star passes behind the larger (2), and in front of the larger (4).

Pulsating variables are stars that vary in light output because they are expanding and contracting in a periodic fashion. Classic examples are the **Cepheid variables** (named after the star Delta Cephei, in the constellation Cepheus), which are yellowish giants and supergiants that increase and decrease in brightness over periods ranging from about 1 day to 80 days. As American astronomer Henrietta Leavitt discovered in 1912, their periods of variation are related to their luminosities – the greater the luminosity, the longer the period. As we shall see in Chapters 8 and 9, this relationship played a pivotal role in measuring the scale of the universe.

Eruptive and cataclysmic variables are stars that brighten and fade abruptly. Among the several varieties are **flare stars**, – cool, low-luminosity stars located towards the bottom of the main sequence. They undergo frequent outbursts of a few minutes' duration, during which they may brighten by several magnitudes. The mechanism responsible for these sudden brightenings is believed to be similar to solar flares, although generally more powerful.

A **nova** is a star that flares up much more dramatically, becoming, at its peak, between a few thousand and a million times as bright as it was prior to the outburst. The rise in brilliancy may take as little as a few hours; then, over a period of months or years, the star gradually fades back to its pre-outburst brightness. The term 'nova', which implies that the star is 'new', is something of a misnomer. In pre-telescopic times, however, each nova did indeed appear to be a 'new' star that appeared where none had been seen before (before the outburst, the star would have been too faint to be visible to the naked eye). A nova seems to be the result of events that occur on the surface of a compact white dwarf that is a member of a close binary system where the two stars are so close together than gas can flow from the companion star to the white dwarf. When enough material has accumulated on the white dwarf's surface it undergoes a runaway thermonuclear reaction that triggers a violent explosion. The dramatic detonation blows surface material into space, leaving the underlying white dwarf relatively unscathed.

A **supernova** is the most dramatic of stellar explosions. It is a true stellar catastrophe, in which a star blows itself apart and, for a few days, may outshine the entire galaxy within which it is embedded. There are two principal types of supernova – Type I and Type II; Type I events can become as brilliant as 10 billion suns whereas Type II events are less brilliant, on average, by a factor of 10 (the mechanisms that give rise to both types are discussed in Chapter 7). On average, a few (perhaps two or three) supernovae per century occur in galaxies similar to our own, but no supernovae have been recorded in our own galaxy since 1604. The most recent naked-eye supernova flared up in 1987. Known as SN1897A, it occurred in the Large Magellanic Cloud, one of our galaxy's closest neighbour galaxies. Despite being about 160,000 light years away, it attained a visual magnitude of 2.8 and so was readily visible to the naked eye (Fig. 6.11).

Table 6.1 Spectral classification of stars.

Class	Effective surface temperature (K)	Colour	Major spectral lines
O	>30,000	blue	mainly singly-ionized and neutral helium and ionized metals; hydrogen lines weak
B	10,000–30,000	blue–white	neutral He lines dominate; hydrogen lines stronger; singly-ionized magnesium and silicon lines
A	7,500–10,000	white	H lines strongest; many lines due to ionized metals (iron, silicon, magnesium, calcium)
F	6,000–7,500	yellow-white	H lines weakening; singly-ionized calcium stronger; neutral metal lines increasing
G	5,200–6,000	yellow	Singly-ionized calcium lines strongest; hydrogen lines weakening further; neutral metals stronger
K	3,700–5,200	orange	neutral metals very strong; hydrogen lines very weak; molecular bands of TiO start to appear
M	2,400–3,700	red	neutral metals very strong; H lines very weak; molecular bands prominent; TiO bands dominate

Figure 6.11 Supernova 1987A (the very bright star just right of centre) was clearly visible to the unaided eye despite its distance of more than 160,000 light-years. Image credit: ESO.

7 Interstellar clouds and the birth, life and death of stars

The space between the stars contains an exceedingly tenuous mix of gas – predominantly hydrogen – and dust (tiny solid grains). The average density of this **interstellar matter** in our galaxy is equivalent to about one hydrogen atom per cubic centimetre of space, which corresponds to one thousand trillion trillionth of the density of air at the surface of the Earth. This material is distributed in a clumpy fashion, much of it being aggregated into clouds, some of which are thousands, or even millions, of times denser than the more thinly spread material that lies between the clouds. These clouds are the ones within which stars are born.

Interstellar gas and emission nebulae

At visible wavelengths, the most obvious evidence for the existence of interstellar gas clouds is provided by **emission nebulae** (singular, **nebula** – the Latin word for cloud) – clouds of gas that shine because embedded within them are one or more hot, highly luminous stars. These stars radiate intense ultraviolet radiation, which is sufficiently energetic to knock electrons off atoms to produce ions. When an electron is recaptured by an ion it usually enters a high-energy orbit round the atomic nucleus, then drops down in a series of steps, to the lowest-energy orbit (*see* Chapter 6). Each downward step results

in the emission of a photon with a particular energy and wavelength. Consequently, the nebula emits light at a number of specific individual wavelengths and its spectrum consists of a series of bright **emission lines** – very different in appearance from the continuous spectrum emitted by a star.

The best-known emission nebula is the Orion nebula, M42 (number 42 in the catalogue of nebulous objects compiled in 1781 by the French astronomer Charles Messier). Located in the 'sword' of Orion, just below the three conspicuous stars that make up Orion's belt, this cloud can be seen – under good dark-sky conditions – with the unaided eye, and is readily visible in binoculars. Embedded within the glowing nebula is a compact group of four hot stars, called the Trapezium, which is responsible for causing it to shine. The visible nebula, which lies at a distance of some 1350 light years from Earth, is about 24 light years in diameter (Fig. 7.1).

Gas clouds reveal themselves in more subtle ways, too. For example, if light from background stars passes through intervening gas clouds while en route to the Earth, those clouds will absorb light at certain particular wavelengths, superimposing additional dark lines on the spectra of stars. **Interstellar clouds** also reveal their presence through the emission of radio and microwave radiation.

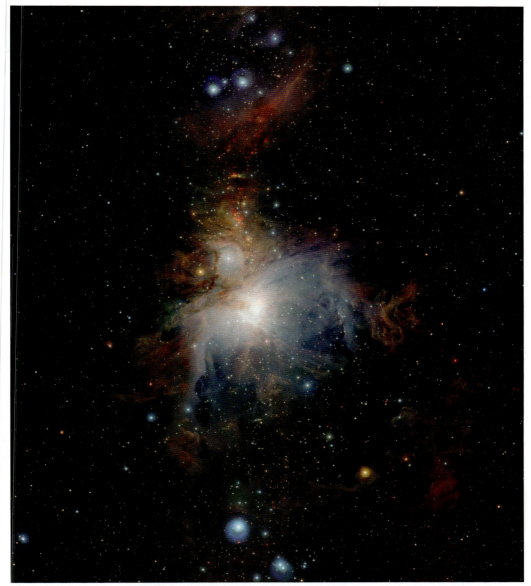

Figure 7.1 The Orion Nebula (M42), which glows because it contains four extremely massive, hot and highly luminous young stars, is the nearest major star-forming 'factory'. Image credit: ESO/J. Emerson/VISTA. Acknowledgement: Cambridge Astronomical Survey Unit.

For example, neutral hydrogen gas – the most abundant element in the universe – radiates at a wavelength of 21.1 cm, and a frequency of 1420 MHz, and molecules (combination of two or more atoms) also emit or absorb radiation at various wavelengths in the infrared, microwave and radio regions of the spectrum. Apart from molecular hydrogen (H_2), most of the molecules that have been identified in interstellar clouds are organic in nature, consisting of carbon combined with elements such as hydrogen, oxygen and nitrogen. Examples include carbon monoxide (CO), formaldehyde (H_2CO), and ethyl alcohol (CH_3CH_2OH).

Interstellar dust

Interstellar clouds also contain tiny grains of solid matter, known as **interstellar dust**. These particles are typically less than a micrometre (a millionth of a metre, or a thousandth of a millimetre) across, comparable with the particles that make up cigarette smoke. If the cloud contains enough dust to blot out most of the light coming from more distant objects which lie behind it, it will show up as a dark **absorption nebula** (or **dark nebula**) – a dark patch silhouetted against the background of stars and luminous nebulae (Fig. 7.2). A classic example is the Coal Sack – a near-circular dark

Figure 7.2 This dark cloud, called Barnard 68, is completely opaque to visible and near-infrared light because of the obscuring effect of the dust particles in its interior. Image credit: ESO.

cloud in Crux Australis (the Southern Cross), which is readily visible to naked eye observers in the southern hemisphere (*see* Fig. 2.1).

Dust grains can also give rise to localized small areas of bright nebulosity. Where light from one or more stars is scattered back towards the Earth by a neighbouring dust cloud, the star, or stars, will appear to be surrounded by a bluish patch of hazy light, which is called a **reflection nebula**.

Starbirth and stellar nurseries

The process of starbirth begins when an individual cloud of gas and dust – or a dense clump (or 'core') within a larger cloud – starts to collapse under its own weight. Over tens of thousands of years, the shrinking cloud becomes hotter and denser). Eventually, when the central temperature rises to about 10,000,000 K, nuclear fusion reactions that convert hydrogen into helium commence,

Figure 7.3 The Lagoon Nebula, a giant cloud of gas and dust, is creating intensely bright young stars, the radiation from which sculpts the distribution of wispy glowing nebulosity and dark dust-laden regions. Image credit: ESO/VPHAS+ team.

the pressure inside the cloud becomes sufficiently great to prevent it from contracting any further, and it becomes a fully-fledged star. Thereafter, the combined effect of a strong **stellar wind** and the young star's powerful radiation output blows away the remaining gas and dust, enabling the newly created star to come into view.

The most favourable conditions for star formation occur in dust-laden **molecular clouds** – clouds of hydrogen and helium that are rich in molecules – where the temperatures are low (typically 10–30 K) and the densities are hundreds or thousands of times greater than the overall average density of interstellar gas. Star formation tends to occur in batches, where a number of dense cores form within a molecular cloud (Fig. 7.4). The best-known and nearest stellar nursery is the Orion molecular cloud, an extensive complex of dust-laden clouds of which the familiar visible Orion nebula is a relatively small part.

The formation of planetary systems

There is strong evidence, for example, from the presence of dusty discs surrounding newly formed stars, and from the fact that several thousand planets have already been

Figure 7.4 The star-forming region NGC 3603 contains a massive cluster of young stars (centre), formed about a million years ago, which is carving out a huge cavity in the gas to its right. Image credit: NASA, ESA and the Hubble Heritage (STScI/AURA)-ESA/Hubble Collaboration.

Figure 7.5 Artist's impression of a newly formed star that is still surrounded by a protoplanetary disc in which planets are forming. Image credit: ESO/L. Calçada.

discovered orbiting around stars, that **planetary systems** are commonplace (*see* Chapter 10). The formation of planetary systems seems to be a natural by-product of the process of star formation, planets forming within flattened discs of gas and dust that surround newborn stars before the remaining gas and dust is blown away (Fig. 7.5).

In the case of our own Solar System, there is a broad consensus that the terrestrial planets and the rocky-metallic cores of the gas-rich giant planets formed by **accretion** – the progressive aggregation of smaller particles and bodies into larger ones. This process is believed to have produced a population of billions of **planetesimals**, irregular solid bodies about 5–10 kilometres in size, and collisions between them led to the growth of some and the break-up of others. When one planetesimal grew to be more massive than its neighbours, it quickly mopped up the others in a process that led to the formation of bodies the size of the Moon or the planet Mercury. Subsequent collisions between these bodies assembled the terrestrial planets and the cores of the giant planets. At the distances at which the giant planets formed, the temperature was low enough for water ice, and other icy materials, to condense into solid particles. This increased the amount of solid material that went into forming their cores, and provided them with enough gravitational influence to trap huge gaseous envelopes.

Approaching the main sequence

While a contracting **protostar** (the precursor of a star) continues to accrete mass from the surrounding cloud, its luminosity rises to a value considerably higher than the luminosity it will have when it eventually becomes a fully-fledged star. Thereafter, it continues to contract without much change in its surface temperature, and its observed luminosity decreases. Plotted on the H-R diagram, it would appear initially above and to the right of the **main sequence** and would then descend towards it. When nuclear fusion reactions switch on and the pressure inside the star becomes sufficiently high to prevent it contracting further, the youthful star settles down at a point on the main sequence which is determined by its mass: the more massive the star, the hotter and more luminous it will be. For example, a main-sequence star ten times as massive as the Sun will be several thousand times more luminous, whereas a star with one-tenth of the Sun's mass will have a luminosity not much greater than a ten-thousandth of the Sun's output.

Protostars with masses of less than about 0.08 solar masses (about 80 times the mass of the planet Jupiter) never attain high enough central temperatures for sustained hydrogen fusion reactions to kick in, and they do not become 'genuine' stars at all. Instead they become **brown dwarfs** – dense gaseous bodies, less luminous than the dimmest main-sequence stars and with surface temperatures that range from about 2,200 K down to 500 K, or even less. They shine only because they are gradually radiating away stored heat that was built up during their early contraction phase.

Main-sequence stars

A **main-sequence star** is in a state of balance, with the pressure exerted by the hot gas in its interior balancing the gravitational self-attraction that is trying to make it contract. Likewise, the rate at which energy is radiated away from its surface is exactly equal to the rate at which energy is being generated in its core. If the output of energy were to increase, the star would expand until, once again, pressure and gravity were in balance; conversely, if the energy generation were to reduce, the star would contract until a state of balance was achieved.

As '**hydrogen burning**' (the conversion of hydrogen to helium by fusion) continues, the proportion of helium in the core increases and the proportion of hydrogen decreases until, eventually, practically all of the hydrogen in the core has been used up. Hydrogen burning in the core then comes to an end, and the core can no longer support the weight of the star. The time that a star spends in its main-sequence stage depends on its mass but, perhaps paradoxically at first glance, the greater its mass, the shorter its lifetime. The reason for this is that high-mass stars are so much more luminous than low-mass ones that – despite having more hydrogen 'fuel' in their cores to start with – they consume fuel so much more rapidly that their supplies quickly run out. The Sun has enough fuel to sustain itself for about 10 billion years (and is currently roughly halfway through its main-sequence stage). The highest-mass stars have lifetimes of only a few million years, but the least massive ones will outlive the Sun many times over.

Post-main sequence evolution

When a star's core runs out of hydrogen, it begins to contract, heating up as it does so. The temperature then rises sufficiently, in a shell surrounding the original core, for hydrogen fusion to commence there. As the core continues to contract, the temperature in the shell rises further and the reactions proceed more rapidly, increasing the star's luminosity and causing its outer envelope to expand. As the star expands, its surface area increases in proportion to the square of the radius (for example, if the radius is doubled, the surface area increases by a factor of four) with the result that, even although the total output of energy is increasing, the amount of energy radiated from each square metre of its surface – and hence the temperature of its surface – decreases. The point on the H-R diagram that represents the star moves upwards (in the direction of increasing luminosity) and to the right (in the direction of decreasing surface temperature), eventually taking the star into the **red giant** region (Fig. 7.6).

While the star is expanding to become a red giant, helium produced in the hydrogen-burning shell continues to be added to the core, which continues to shrink, becoming progressively hotter. Eventually, when the core temperature reaches about 100,000,000 K, a helium-burning reaction, called the **triple-alpha reaction**, commences. This reaction welds together three helium nuclei (which are sometimes known as alpha particles – hence the name of the reaction) to form a nucleus of carbon, releasing energy in the process. In

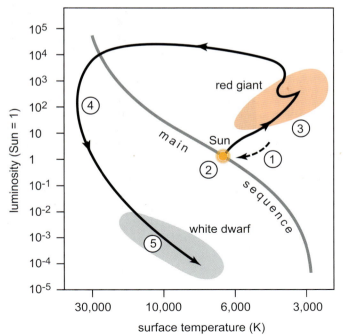

Figure 7.6 Evolution of a star like the Sun. Plotted on the H-R diagram, a contracting protostar (1) settles down on the main sequence (2). When it has consumed the hydrogen fuel in its core, it expands to become a red giant (3), after which it sheds its outer layers to form a planetary nebula, and then (4) it shrinks and fades to become a white dwarf (5).

addition, some of the carbon produced in the triple-alpha reaction can interact with helium nuclei to form oxygen. Once the triple-alpha reaction has become established in the star's core, the star reaches a fairly stable state again, and settles down for a time as a distended and highly luminous red giant.

Because its luminosity is now so high, the star rapidly consumes its reserves of helium fuel. When the entire core has been converted into carbon, helium burning ceases there, but as the core again starts to contract, helium burning continues for a time in a shell around the now inert core. Because the rate of the triple-alpha reaction is extremely sensitive to temperature, the star can become unstable at this stage and may expel a shell, or shells, of unconsumed hydrogen. With no thermonuclear energy source to support it, the star shrinks rapidly and the temperature at its surface rises to well over 30,000 K, sometimes even as far as 100,000 K. Ultraviolet radiation from its surface ionizes the expanding shell of gas, causing it to fluoresce (emit light), so giving rise to a phenomenon known, somewhat misleadingly, as a **planetary nebula** (Fig. 7.7).

The surface area of the shrinking star decreases dramatically and its luminosity rapidly declines. Plotted on the H-R diagram, the star moves from right to left (cooler to hotter) and rapidly downwards (from high to low luminosity). Because the shrinking core of a solar-mass star cannot contract far enough to raise its temperature high enough for carbon-burning nuclear reactions to commence, the star continues to shrink until the pressure exerted by fast-moving electrons in its interior becomes so great that it ceases to contract. By the time is has reached this

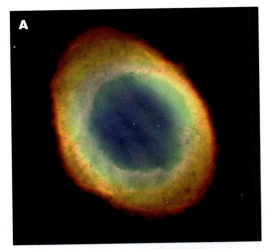

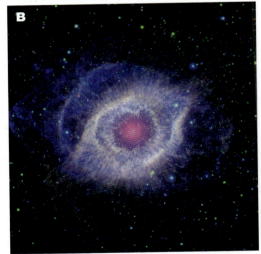

Figure 7.7 Planetary nebulae: **A**, the Ring Nebula (M57) surrounds the intensely hot dying star at its centre. About a light-year in diameter, it lies some 2000 light years distant in the direction of the constellation Lyra. Image credit: The Hubble Heritage Team (AURA/STScI/NASA). **B**, This composite infrared/ultraviolet image of the Helix nebula shows several expanding shells of gas that were expelled by the dying star before its shrinking core became a white dwarf. Image credit: NASA/JPL-Caltech.

state, the dying star has shrunk to about one-hundredth of the radius of the Sun (about the size of the Earth) and one-thousandth of its luminosity. It has become a **white dwarf**. Composed mainly of carbon and oxygen, a white dwarf is so dense (about a million times denser than water) that a teaspoonful of its material, if brought back to the Earth, would weigh about a tonne.

Unable to generate energy by means of nuclear reactions, the dying star gradually cools and fades. Over many billions of years it will eventually become a cold, dark body called a **black dwarf**; however, this process takes so long that there has not yet been enough time since the origin of the universe for any white dwarf to become a black dwarf.

In five or six billion years' time, the Sun will expand to between 20 and 100 times its present size, its luminosity will increase a thousand-fold, and its surface temperature will drop to around 3000 K. By the time it becomes a red giant, the planet Mercury – and possibly even Venus – will be swallowed up, and the temperature at the Earth's surface will have risen to about 1500 K. The oceans will evaporate, the atmosphere will be driven off into space, and life on the surface of our planet will become impossible. By the time the Sun has become a white dwarf, the Earth, or what remains of it, will have cooled to a frigid husk, and life as we know it will no longer be possible anywhere in the Solar System.

High-mass stars

Stars comparable to, or less massive than, the Sun will evolve in a similar fashion, but for stars of significantly higher mass, the late stages of evolution proceed differently and much more rapidly. For example, a star of around five solar masses will leave the main sequence after about 100 million years. After reaching the red giant stage, its outer layers will become unstable and it will become a pulsating variable such as a Cepheid or a long-period variable.

The carbon and oxygen core of a high-mass star contracts and heats up further. If the mass of the star exceeds about four solar masses, its core temperature can rise to around 600,000,000 K, at which point carbon-burning reactions commence, these producing end products such as neon, magnesium, oxygen and more helium. Heavier stars develop even hotter cores, within which fusion produces heavier elements such as silicon. In very high-mass stars, where the core temperature can reach about three billion kelvins, successive fusion reactions lead to the formation of iron. In order to sustain a star, nuclear fusion reactions must release energy. But in order to make heavier elements from iron, energy would have to be put in. Consequently, once a star's core has been converted into iron, it can no longer be supported by fusion reactions and, deprived of support, it abruptly collapses.

If the core of a collapsing star exceeds about 1.4 solar masses – a limit that is called the Chandrasekhar limit after the Indian astrophysicist who first computed it – it cannot become a white dwarf, because gravitational forces will overwhelm the pressure exerted by fast-moving electrons. A fraction of a second after the onset of **core collapse**, the temperature becomes so high that atomic nuclei break up into their constituent components. The density of the collapsing core rapidly becomes so great that electrons combine with protons to form electrically neutral neutrons, which can be packed very close together indeed. Within a few tenths of a second, the core shrinks to a radius of about 10 km, its density rises to around 4×10^{17} kg/m³ (about 400 trillion times the density of water) and, provided that its mass is less than two to three times the mass of the Sun, the pressure exerted by close-packed neutrons will halt the collapse. The end product will be a **neutron star**, which is so dense that if a teaspoonful of its material were brought back to the Earth and placed on the long-suffering bathroom scales, it would weigh about 400 million tons!

The rest of the star falls in on top of its collapsed core, and rebounds, sending a shock wave surging outward, which blasts most of the star's material into space in a cataclysmic explosion that is called a **supernova**. Within a day or two the shattered star's luminosity can rise to 600 million solar luminosities (absolute magnitude –17) – brighter than a small galaxy. During the explosion, and subsequently through nuclear processes that take place in the expanding cloud of debris, a wide range of chemical elements – including those that are heavier than iron – are produced and scattered into space. What is left behind is a neutron star surrounded by an expanding cloud of material that is called a **supernova remnant**. As the supernova remnant merges into neighbouring gas clouds, it seeds them with heavier elements, so ensuring that subsequent generations of stars are born within clouds that have been enriched with a higher proportion of heavier elements than the clouds from which earlier generations of stars were formed. Most of the elements from which planet Earth, and we ourselves, were formed were forged in supernova explosions.

The best-known supernova remnant is the Crab Nebula (M1) in the constellation Taurus, the remains of a supernova that was seen by Chinese observers in AD 1054. Despite being located at a distance of 6500 light-years, at peak brilliancy this supernova was visible to the naked eye in a clear blue daylight sky; it subsequently faded from view in a few months. This turbulent remnant is currently 11 light years in diameter and is a source of all kinds of radiation from x-rays to radio waves (Fig. 7.8).

The destruction of a high-mass star in this way gives rise to a **Type II supernova** (or 'core-collapse' supernova). A completely different mechanism is responsible for producing another kind of supernova (Type Ia), which is at least ten times more luminous. A **Type Ia supernova** is believed to occur when carbon fusion reactions are abruptly triggered deep inside a white dwarf. This seems likely to occur when a white dwarf with a mass very close to the Chandrasekhar limit is a member of a close binary system and is accreting a stream of material from its companion. As the accreted hydrogen burns to helium, it adds to the mass of the white dwarf, pushing it ever closer to the limit. When it reaches that limit, the star begins to collapse, igniting the carbon and oxygen inside and triggering a catastrophic blast of energy that blows the star completely to pieces, leaving no central remnant at all.

Neutron stars and pulsars

In 1967, Jocelyn Bell (now Professor Bell-Burnell), then a research student working with Antony Hewish at Cambridge, discovered a curious source in the sky that emitted a short pulse of radio waves at intervals of 1.33

Figure 7.8 Supernova remnants: **A**, this composite x-ray, visible and infrared image of the Crab Nebula shows expanding filaments of debris from the explosion which created the nebula and the neutron star at its heart. Image credit: X-ray: NASA/CXC/SAO/F. Seward; Optical: NASA/ESA/ASU/J. Hester & A. Loll; Infrared: NASA/JPL-Caltech/Univ. Minn./R. Gehrz. **B**, Kepler's supernova remnant is the debris from a Type Ia supernova that was seen by Kepler in 1604, and which was caused by the destruction of a white dwarf. Image credit: X-ray: NASA/CXC/NCSU/M. Burkley et al.; Optical: DSS.

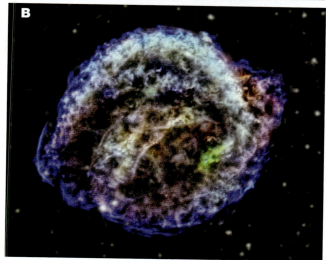

seconds. Within a few months, the Cambridge team had discovered several similar objects, all of which maintained their pulse periods with astonishing regularity. These pulsating radio sources came to be known as **pulsars**. It was soon realized that this curious phenomenon could best be explained by assuming that these sources were rapidly rotating neutron stars.

All stars rotate. If a star were to collapse down to the size of a neutron star, it would end up spinning very rapidly indeed – tens

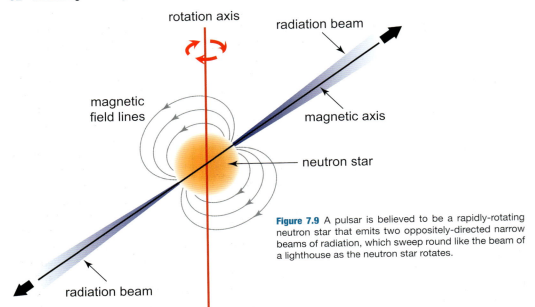

rotation axis

radiation beam

magnetic
field lines

magnetic axis

neutron star

radiation beam

Figure 7.9 A pulsar is believed to be a rapidly-rotating neutron star that emits two oppositely-directed narrow beams of radiation, which sweep round like the beam of a lighthouse as the neutron star rotates.

or hundreds of times a second. The observed periods of the known pulsars range from about eight seconds down to about a millisecond. Only a tiny, dense neutron star could spin this fast and hold itself together. The magnetic field at the surface of a collapsing star grows rapidly in strength, and at the surface of a neutron star is likely to be somewhere between a hundred million and ten trillion times the magnetic field strength at the surface of the Earth. Charged particles accelerated by magnetic fields emit electromagnetic radiation and, because magnetic field lines are bunched together at the north and south magnetic poles of a neutron star, the emitted radiation will be concentrated into two narrow beams directed along the magnetic axis, one from the north and the other from the south magnetic pole. If the

magnetic axis is tilted at an angle to the rotation axis, the neutron star's rapid rotation will cause its beams of radiation to sweep round like the beam of a lighthouse. Each time the beam points in our direction, we receive a pulse of radiation (Fig. 7.9).

Strong confirmatory evidence that pulsars are indeed rapidly rotating neutron stars and that neutron stars are created in core collapse supernovae is provided by the fact that pulsars have been discovered located within a number of supernova remnants, notably the Crab nebula.

Black holes

A **black hole** is a region of space into which matter has fallen, and within which gravity is so powerful that nothing – not even light itself – can escape to the outside universe. A black hole will be created when a sufficiently large mass is compressed within a sufficiently small radius – a situation that is likely to occur when a very high-mass star runs out of nuclear fuel and collapses under its own weight.

The basic concept of a black hole was first suggested as far back as the late eighteenth century when English natural philosopher John Michell (in 1783) and French mathematician Pierre Simon de Laplace (in 1796) independently suggested that bodies might exist which were so massive that their escape velocities (*see* Chapter 3) would be greater than the speed of light. Thinking of light as a stream of tiny particles, they argued that particles of light would be unable to escape from such bodies. The modern theory of black holes is based on Einstein's theory of gravity (General Relativity) rather than Newton's, but it leads to the same general conclusion, that if a particular mass of material is compressed within a sufficiently small radius (the **Schwarzschild radius**), light will be unable to escape from within that radius. The size of the Schwarzschild radius, in kilometres, is equal to about $3 \times$ (the mass of the body expressed in solar masses); for the Sun, the Schwarzschild radius is just 3 km. Although there is no natural process in the present-day universe that could compress the Sun sufficiently to turn it into a black hole, the situation can be completely different for stars that are much more massive.

If the mass of the collapsing core of a very high-mass star exceeds the maximum permitted mass of a neutron star (between two and three solar masses), then nothing can halt its collapse, and it will continue to fall in on itself until all of its material is compressed into a point of infinite density, which is called a **singularity**. Before this happens, the collapsing star will pass inside its own Schwarzschild radius, disappear from view, and create a black hole. The outer parts of the star may be blasted away in a supernova explosion (or an even more extreme event called a **hypernova**), or may instead fall directly into the newly created black hole. The boundary of a black hole is called the **event horizon** because no information about any event that might occur inside that 'horizon' can be communicated to the outside Universe.

Although it cannot be seen directly, a black hole can reveal its presence through its gravitational influence. For example, if a black hole is a member of a binary system, astronomers will see a visible star revolving around an invisible object, and if the mass of the invisible object is more than about three solar masses (the upper limit for a neutron star) then there is a good case for thinking that a black hole is involved. If a black hole and its companion star are sufficiently close to each other, the gravitational pull of the black hole will drag a stream of gas from the star. Because of its orbital motion, this gas will form a rapidly rotating disc (an **accretion disc**) around the black hole. Infalling material ploughing into this disc raises its temperature to millions or even hundreds of millions of kelvins, causing it to emit x-rays. If an x-ray source in the sky coincides with a binary that includes an invisible companion with a mass greater than the maximum limit for a neutron star, the case for a black hole being present is a strong one. Many examples of this kind of object have been detected since the first – an object called Cygnus X-1 – was identified in 1972 (Fig.7.10).

Figure 7.10 Artist's impression of Cygnus X-1, a black hole which is dragging material from a companion star; as the gas spirals in, it is heated to very high temperatures and radiates x-rays. Image credit: NASA, ESA, Martin Kornmesser (ESA/Hubble).

Summing up the fates of stars

The overwhelming majority of stars will eventually run out of nuclear fuel and will then shrink to form compact white dwarfs, which over many hundreds of billions of years, will cool down to become dark, black dwarfs. Stars with initial masses greater than about eight solar masses will collapse to form highly compressed neutron stars, and their gaseous envelopes will be blasted into space in supernova explosions. The most massive stars of all seem destined to collapse without limit to form those most enigmatic of astrophysical objects – black holes.

8   Galaxies

On a clear dark night, when the Moon is not around, it is easy to see the **Milky Way**, a faint band of misty light that stretches across the sky from horizon to horizon. It extends through many well-known constellations, stretching northwards through Centaurus, Crux Australis, Vela and Canis Major, between Orion and Gemini, and on through Auriga, Perseus and Cassiopeia, before heading southwards again through Cygnus and Aquila to the rich star fields of Scorpius and Sagittarius. Although humankind had been aware of its existence since the earliest of times, it was not until the beginning of the seventeenth century that astronomers such as Galileo were able to use telescopes to reveal that it is made up of the combined light of innumerable stars. A century and a half later, in 1759, Thomas Wright suggested that stars were distributed in a vast flattened disc and that when we look along the plane of this disc we see the great concentration of stars that makes up the Milky Way (Fig. 8.1).

The Milky Way galaxy

We now know that the Milky Way is a **galaxy**, a disc-shaped aggregation of stars, gas, and dust, with a diameter of more than 100,000

Figure 8.1 The Milky Way above the domes of La Silla Observatory, Chile. The central part of the Milky Way is visible behind the distant domes (lower left); the two Magellanic Clouds can be seen on the right. Image credit: ESO/Z. Bardon (www.bardon.cz) ProjectSoft (www.projectsoft.cz).

light years, which contains several hundred billion stars. The Sun is located about 27,000 light years away from the centre of the system (the galactic centre) – just over halfway from the centre to the edge. The central **nucleus** of the galaxy is surrounded by an ellipsoidal distribution of relatively close-packed stars (the **central bulge**, or **nuclear bulge**), which measures about 15–20 thousand light years in diameter and about 6000 light years in 'height'. So much dust lies between the Earth and the galactic centre that it cannot be seen directly at visible wavelengths. It can, however, be studied at infrared and radio wavelengths, which penetrate the intervening dusty clouds.

The central bulge takes the form of an elongated bar, which is almost end-on as viewed from the present location of the Solar System. Whereas the central bulge is dominated by old red giants, the disc contains most of the gas, dust, star-forming regions and highly luminous hot young stars, together with thousands of relatively young star clusters. Surrounding the bulge, and extending in a near-spherical distribution above, below and beyond the disc is the **halo**. This is composed of about two hundred **globular clusters** (near-spherical clusters that contain from around ten thousand to upwards of a million member stars) and a thinly scattered distribution of old stars. Globular clusters contain very little gas and dust, consist predominantly of old stars, and are some of the most ancient structures in our galaxy. Two fine examples are M13 in the constellation Hercules, and Omega (ω) Centauri, the brightest one, which can be seen with the naked eye (Fig. 8.2).

The entire galaxy is rotating, with each star following its own individual orbit around the galactic centre. The Sun, together with the Solar System, is moving along its orbit round that galactic centre at a speed of about 250 km/s and takes about 225 million years to complete each circuit.

If most of the mass in the galaxy were concentrated where most of the stars appear to be – in and around the nuclear bulge – then stars and gas clouds beyond the central regions would behave rather like planets in their orbits round the Sun: their orbital speeds should decrease with increasing distance. In fact, studies of the motion of stars, and of gas clouds that can be detected out to distances beyond the visible disc of stars, show that their orbital speeds remain much the same, or even increase, with increasing distance from the centre. Because the rotational velocity

Figure 8.2 Omega Centauri – the largest and brightest globular cluster in the sky – contains about 300,000 stars. Image credit: ESO/INAF-VST/OmegaCAM. Acknowledgement: A. Grado/INAF-Capodimonte Observatory.

at a particular distance depends on the total amount of mass contained within that radius, these observations show that our galaxy must contain 5–10 times as much non-luminous matter as visible matter, and that this **dark matter** is distributed in a halo that extends well beyond the visible parts of the galaxy to a radius of 200,000 light years or more. The nature of this matter is a question of much debate, and will be explored in Chapter 9.

Within the disc, highly luminous young stars, clouds of gas and dust, and star-forming regions are concentrated into curved 'arms' (**spiral arms**) that appear to spread outwards from the central bulge in a spiral pattern (Fig. 8.3). Our galaxy's spiral structure

Figure 8.3 This artist's impression of the Milky Way galaxy shows the principal spiral arms, the central bar and the location of the Sun. Image credit: NASA/JPL-Caltech/ESO/R. Hurt.

comprises several major arms and a number of shorter segments, one of which – the 'Orion spur' – contains the Solar System and the star-forming regions of Orion. Although there is still considerable uncertainty about the details, the four major arms are the Norma arm, which emanates from the near side of the central bar, passes behind the galactic nucleus and re-emerges as the 'outer arm'; the Perseus arm, which also passes behind the nucleus, curves round and passes outside the Sun's location; and the Carina-Sagittarius and Crux-Scutum arms, which pass between the Sun and the galactic centre.

A compact source of radio and shorter-wave emission, called Sagittarius A*, appears to mark the very centre of the galaxy. Sagittarius A* has an angular diameter of less than 0.00004 arcsec, which corresponds to a physical diameter of about 0.3 AU, equivalent to less than one-third of the radius of the Earth's orbit around the Sun. Doppler measurements show that stars and clouds of ionized gas are revolving rapidly round the galactic centre, which implies that they must be subject to the gravitational influence of a very compact massive object, which is generally believed to be a **supermassive black hole** with a mass of at least four million solar masses. Accretion of gas and shredded stars onto this black hole is thought to be the underlying source of the energy radiated by Sagittarius A*.

Galaxies

Until the 1920s, there was an ongoing, and sometimes heated, debate among astronomers as to whether the spiral-shaped 'extragalactic nebulae' (as they were called at the time) that they could see above and below the plane of the Milky Way were independent star systems far beyond our own, or small appendages to our own galaxy. In order to resolve this debate, their distances had to measured – a feat that was first accomplished in 1923 by American astronomer Edwin Hubble. His technique depended upon a relationship between the luminosities of Cepheid variable stars (*see* Chapter 6) and their periods of variation, which had been discovered in 1912 by fellow American astronomer, Henrietta Leavitt: the more luminous the Cepheid, the longer its period. Hubble managed to identify a number of Cepheid variables embedded within what, at that time, was called the Andromeda nebula; he measured their periods and used the **period–luminosity relationship** to determine their true luminosities. By comparing their measured apparent brightnesses with their luminosities, and knowing that the apparent brightness of a star decreases in proportion to its distance squared, he was able to calculate their distances (Fig. 8.4). Hubble's observations indicated that the Andromeda 'nebula' lay far beyond the confines of the Milky Way, and showed that it was indeed a separate galaxy in its own right. Current measurements indicate that the Andromeda galaxy, as it is now known, lies at a distance of 2,500,000 light years; it is a spiral system similar to our own, but with an overall diameter about half as great again as the Milky Way system.

Following Hubble's ground-breaking measurement of the Andromeda galaxy's distance, subsequent observations soon confirmed that other so-called 'extragalactic nebulae' were independent galaxies, too, and we are now very well aware that the Milky Way galaxy is just one of the many billions of galaxies that populate the observable universe.

Figure 8.4 The arrow indicates the position of the first Cepheid variable star that Edwin Hubble identified in the Andromeda galaxy in 1923; the four inset images show the varying brightness of this star. Image credit: NASA, ESA, and the Hubble Heritage Team (STScI/AURA).

The distances of the galaxies

There are two fundamental approaches to measuring the distances of galaxies: 'standard candles' and 'standard rulers', both of which rely on astronomers being able to identify individual objects within galaxies, the inherent luminosity or physical diameter of which can be estimated with some degree of confidence. The **standard candle** approach relies on the fact that (provided no light is absorbed in the intervening space) the apparent brightness of a distant object is inversely proportional to the square of its distance. Astronomers, therefore, can determine the distances of galaxies by comparing the observed apparent brightness of objects, such as Cepheid variables, bright supergiant stars, globular clusters, novae or supernovae, with assumed values for their luminosities. The **standard ruler** approach relies upon identifying individual objects such as large emission nebulae, planetary nebulae and globular clusters, and comparing their measured apparent sizes with assumed values for their diameters. Because apparent angular diameter is inversely proportional to distance (if the distance is doubled, the apparent diameter is reduced to one half) it is straightforward, in principle, to calculate the distance of the object, and hence the distance of the galaxy within which it is embedded. Where possible, a variety of different distance indicators are used in order to arrive at a weighted average for the distance of an individual galaxy.

Classifying galaxies

Galaxies display a wide variety of shapes, sizes and compositions. The **Hubble classification scheme** identifies four basic types – spiral, barred spiral, elliptical and irregular – and arranges them in a sequence which, because of its shape, is commonly known as the tuning fork diagram (Fig. 8.5).

Elliptical galaxies are denoted by the letter E followed by a number from 0 to 7 to indicate the degree of flattening of their observed elliptical shapes. An E0 galaxy appears circular, whereas an E7 galaxy is markedly flattened, with its major axis being more than three times its minor axis. Elliptical galaxies tend to have a rather bland appearance, showing little sign of any detailed structure, and usually contain little, or relatively modest amounts of, gas or dust. The great majority of ellipticals are small, low-mass 'dwarf' systems, but giant and supergiant ellipticals are found in the central regions of the more massive clusters of galaxies.

Spiral galaxies, denoted by the letter S, consist of a central nucleus and bulge surrounded by a flattened disc of stars, gas and dust that is organized into a pattern of spiral arms (Fig. 8.6). They are classified according to the size of the central bulge, the relative tightness of the spiral pattern and the degree of clumpiness within their arms. An Sa galaxy has a large central bulge and tightly wound, relatively smooth arms; an Sb galaxy has a somewhat smaller nuclear bulge and less tightly wound arms, which often contain

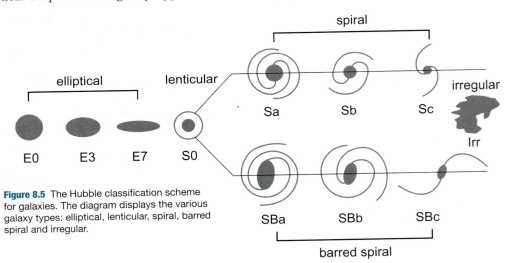

Figure 8.5 The Hubble classification scheme for galaxies. The diagram displays the various galaxy types: elliptical, lenticular, spiral, barred spiral and irregular.

A

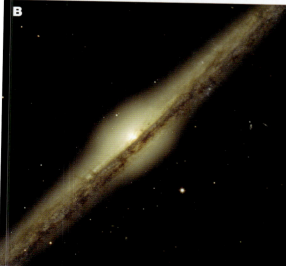

B

Figure 8.6 Spiral galaxies: **A**, the Pinwheel galaxy (M101) is a gigantic spiral, seen 'face-on'; some 170,000 light-years in diameter, it contains about a trillion stars. Image credit: European Space Agency & NASA (full acknowledgements below). **B**, edge-on spiral galaxy, NGC 4565, displays its central bulge and dust lanes in its disc. Image credit: ESO.

Figure 8.6A: European Space Agency & NASA. Acknowledgements: Project Investigators for the original Hubble data: K.D. Kuntz (GSFC), F. Bresolin (University of Hawaii), J. Trauger (JPL), J. Mould (NOAO), and Y.-H. Chu (University of Illinois, Urbana); Image processing: Davide De Martin (ESA/Hubble); CFHT image: Canada-France-Hawaii Telescope/J.C. Cuillandre/Coelum; NOAO image: George Jacoby, Bruce Bohannan, Mark Hanna/NOAO/AURA/NSF.

conspicuous emission nebulae, star-forming regions and clusters of hot young stars; and an Sc galaxy has a relatively small nucleus and open, loosely wound arms usually with a knotty structure, dominated by emission nebulae and youthful clumps of stars. In **barred spiral galaxies** (denoted by SB) the arms emerge from the ends of what looks like a rigid bar (or elongated ellipsoid) of stars and luminous matter that straddles the nucleus (Fig. 8.7) They are classified as SBa, SBb and SBc according to similar criteria to the spiral galaxies. Intermediate categories are labelled appropriately (for example, a galaxy midway in appearance between Sa and Sb would be designated Sab). The Milky Way galaxy used to be regarded as an Sbc spiral, but in more recent times, a strong body of evidence has accumulated which shows that our galaxy's nuclear bulge is elongated – about twice as long as it is broad – and so it may better be classified as an SBbc barred spiral.

Intermediate between spiral and elliptical galaxies are **lenticular** (lens-shaped) **galaxies**, which are denoted by S0 or SB0. Although the nucleus of an S0 galaxy is surrounded by a

Figure 8.7 The barred spiral galaxy, NGC 1300, showing the arms emerging from opposite ends of the bar that straddles its nucleus. Image credit: NASA, ESA, and The Hubble Heritage Team (STScI/AURA).

disc, the disc is rather featureless and contains no indication of spiral structure. **Irregular galaxies**, which have no obvious nucleus or ordered structure, are denoted by Irr.

Compositions and masses

Of the inherently more luminous galaxies, about 75% are spirals or barred spirals and about 20% are elliptical or lenticular. But a large proportion of the elliptical and irregular galaxies are dwarf systems, and when this is taken into account, the overall proportion of spirals reduces to 20–30%.

Galaxy masses range from as little as a million solar masses for the smallest dwarf systems to about ten trillion (10^{13}) solar masses for the supergiant ellipticals, and their total luminosities (the combined light output of their stars and luminous gas clouds) from a few hundred thousand to a trillion times the luminosity of the Sun. Their visible diameters range from a few thousand light years to several hundred thousand light years, and the extended haloes surrounding some of the supergiant ellipticals can be as large as 5 million light years across. Measurements of the speeds at which stars and gas clouds revolve at different distances from the centres of galaxies show that, as with the Milky Way galaxy, up to 90% of the mass of a typical galaxy consists of dark matter, much of which lies in their outer regions.

Groups and clusters of galaxies

Most galaxies are members of **galaxy clusters**, which are collections of galaxies held together by gravity. The most massive clusters contain up to several thousand members. Smaller clusters, which contain up to a few dozen members, are called groups. Numerous collisions and close encounters have taken place (and still are taking place) between member galaxies within clusters, leading to the disruption of some and the growth of others, and triggering intense bouts of star formation and the ejection of heated gas into intergalactic space. The evolution of galaxies, over many billions of years, has been a violent process.

The Milky Way is the second-largest member of a group, containing some 30 to 40 galaxies, which is called the **Local Group**. The three principal members – the Andromeda galaxy, the Milky Way galaxy and the more modest Triangulum galaxy (M33) – are spirals, whereas most of the others are dwarf ellipticals and irregulars; the Andromeda galaxy is the largest and most massive member.

The Milky Way has two substantial satellite galaxies – the Large and Small Magellanic Clouds – both of which are visible to the naked eye in southern hemisphere skies. Located at a distance of 163,000 light years, the **Large Magellanic Cloud** is a substantial galaxy in its own right (Fig. 8.8). With an overall diameter of about 30,000 light years and a population of about ten billion stars, it was originally considered to be an irregular galaxy, but it does contain a central bar structure, and a hint of a truncated spiral arm that contains the Tarantula nebula, a huge glowing gas cloud that surrounds a vigorous region of star formation that was probably triggered by a close encounter with our galaxy a few billion years ago. The **Small Magellanic Cloud** is slightly further away and has about half the diameter and a quarter of the mass of its larger sibling.

The nearest major cluster is the **Virgo cluster** (Fig. 8.9). Located at a distance of about 50 million light years, in the direction

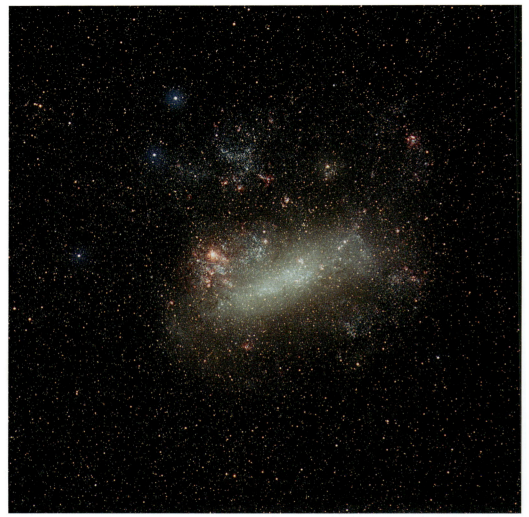

Figure 8.8 This image of the Large Magellanic Cloud shows the huge Tarantula nebula above the left end of the galaxy's central bar. Image credit: Eckhard Slawik (e.slawik@gmx.net).

of the constellation Virgo, it contains about two thousand members, and is centred on the giant elliptical galaxy M87. The Virgo cluster is by no means the largest or most massive cluster within telescopic range; for example, the Coma cluster, which is located at a distance of about 300 million light years, contains up to ten thousand galaxies.

Figure 8.9 The part of the Virgo cluster of galaxies that is shown here contains two massive elliptical galaxies, M84 and M86 (lower half of image). Image credit: NOAO/AURA/NSF.

On an even larger scale, clusters and groups of galaxies are gathered into loose sprawling structures called **superclusters**, which have diameters of around 100 million light years, and which typically contain two or three major clusters and a number of smaller groups. The Local Group lies on the fringe of the Virgo supercluster, which spans 100 million light years and contains about five thousand member galaxies. Overall, the distribution of matter in the universe is rather frothy. Galaxies and intergalactic matter are aggregated into clusters and superclusters, long filaments and sheets, which are separated by voids where few, if any, galaxies are seen. The fact that matter is distributed in this web-like pattern throughout the universe is related to the way in which structure formed in the very early universe (*see* Chapter 9).

Dark matter in galaxy clusters

Within clusters and groups, the speeds at which the individual galaxies move around are such that, were the mass of the group or cluster equal only to the combined mass of its visible constituent stars and gas clouds, there would be far too little net gravitational attraction to prevent cluster members from escaping, and the clusters would have dispersed and lost their identities before now. To hold themselves together, galaxy clusters must contain a great deal more mass, in the form of dark matter, than meets the eye. Further evidence for the existence of huge amounts of dark matter comes from x-ray studies of the exceedingly hot ionized gas that permeates the space between the member galaxies in massive clusters. Once again, the additional gravitational influence of great amounts of dark matter is needed in order that clusters retain this exceedingly hot gas.

More evidence comes from the phenomenon of **gravitational lensing** (Fig. 8.10). According to Einstein's general theory of relativity, the presence of mass distorts or 'curves' space so that a ray of light passing close to a massive body will be deflected by an amount that depends on the mass of the body and the distance at which the light ray passes. A foreground distribution of mass, such as a galaxy cluster, will deflect rays of light coming from more distant sources and will act like a lens to form magnified and distorted images of more distant background galaxies (Fig. 8.11). By analysing the distorted images produced in this way, astronomers can measure the amount and distribution of dark and luminous matter in the foreground cluster. Gravitational lensing has confirmed that galaxy clusters contain from ten to a hundred times as much dark matter as luminous matter.

Active galaxies and quasars

The term **active galaxy** is used to describe galaxies that have unusual characteristics, and often have a peculiar, or disturbed, appearance. Whereas most of the energy emitted by an ordinary galaxy is the light emitted by its constituent stars and gas clouds, an active galaxy radiates strongly across a very wide range of wavelengths, and is much more luminous at x-ray, ultraviolet, infrared and radio wavelengths than a normal galaxy such as our own. Active galaxies typically radiate hundreds or thousands of times

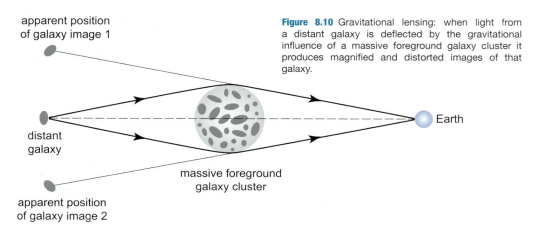

apparent position
of galaxy image 1

distant
galaxy

apparent position
of galaxy image 2

massive foreground
galaxy cluster

Earth

Figure 8.10 Gravitational lensing: when light from a distant galaxy is deflected by the gravitational influence of a massive foreground galaxy cluster it produces magnified and distorted images of that galaxy.

Figure 8.11 In this image of the massive galaxy cluster, Abell 1689, X-ray emission from hundred-million-degree gas is shown in purple. The long arcs among the visible galaxies are caused by gravitational lensing of background galaxies. Image credit: NASA/CXC/MIT/E.-H Peng et al.; Optical: NASA/STScI.

as much energy as the Milky Way galaxy, much of that energy being emitted by charged particles moving at large fractions of the speed of light in magnetic fields. Most significantly, an active galaxy contains a compact, highly luminous core that, in many cases, varies markedly and rapidly in brightness and from which, in many examples, emerge two oppositely directed jets of radiating material. The seat of this activity is called an **active galactic nucleus** (**AGN**). The principal types of active galaxy are radio galaxies, quasars, BL Lacertae objects (or 'blazars') and Seyfert galaxies (Fig. 8.12).

Figure 8.12 Active galaxies: **A**, This composite image of active galaxy, Centaurus A, shows jets and lobes emanating from the vicinity of the black hole at its centre. The x-ray jet (blue) extending towards the upper left is about 13,000 light-years long. Image credit: ESO/WFI (Optical); MPIfR/ESO/APEX/A. Weiss et al. (Submillimetre); NASA/CXC/CfA/R. Kraft et al. (X-ray). **B (opposite)**, Radio galaxy, Hercules A. Radio data reveal huge jets of high-energy particles shooting out from the massive elliptical galaxy at the centre of the image. The separation between the two lobes (radio-emitting clouds) on either side is about 1.5 million light-years. Image credit: NASA, ESA, S.Baum and C. O'Dea (RIT), R.Perley and W. Cotton (NRAO/AUI/NSF), and the Hubble Heritage Team (STScI/AURA). **C (opposite)**, This image of the centre of the massive elliptical galaxy, M87, shows the jet emerging from the central core, which harbours a 2.6 billion solar mass black hole. Image credit: Tod R. Lauer (NOAO), Sandra M. Faber (CSC), C. Roger Lynds (NOAO), and the Wide Field/Planetary Camera Imaging Team.

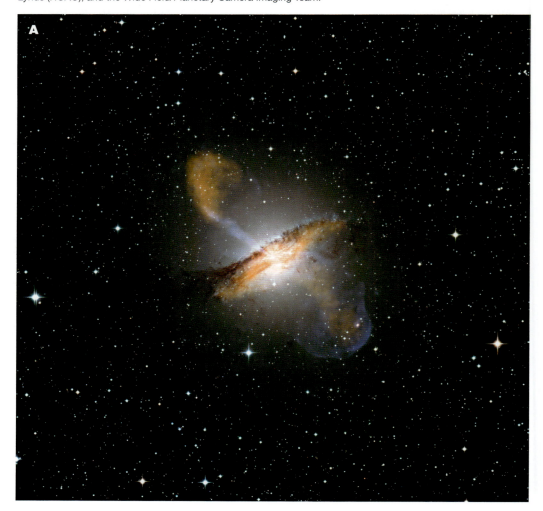

B

C

Radio galaxies are powerful sources of radio emission. Typically, much of the emission comes from two huge clouds, or 'lobes', of material, located beyond and on either side of the underlying visible galaxy. During the early 1960s, some radio sources were found to coincide in position with objects that looked like stars. Consequently, they became known as **quasars** (quasi-stellar radio sources). Although it has since been shown that only about ten percent of these objects is a strong radio emitter, and the term **quasi-stellar object** (**QSO**) was invoked to describe the radio-quiet versions, the term quasar is still widely used to describe both types of object. When quasars were first shown to lie at immense distances, astronomers were at a loss to explain how such an apparently compact and star-like object could radiate the huge quantities of energy (thousands of times more than a conventional galaxy) which their apparent brightnesses implied. However, in the 1980s and 1990s, astronomers found that some of them, at least, were embedded within faint galaxies, and it is now believed that all quasars are in fact extreme examples of active galactic nuclei.

BL Lacertae objects (BL Lacs) are similar in appearance to quasars, but, unlike quasars, do not have any obvious lines in their spectra. Many of them display rapid and dramatic variability, and these objects, together with the most violently variable of the quasars, have come to be known as **blazars**. **Seyfert galaxies** are a distinctive class of spiral and barred spiral galaxies – first identified in 1943 by Carl Seyfert – which have a bright, compact nucleus. Although less luminous than quasars, Seyferts are brighter than conventional galaxies, especially at infrared, ultraviolet and x-ray wavelengths.

The energy machine in active galaxies

The widely held view is that every active galactic nucleus contains a supermassive black hole with a mass in the region of 10 million to several billion solar masses. Because galaxies rotate, matter falling in towards the central black hole will conserve angular momentum and form a rapidly rotating disc, or doughnut-shaped torus, of material encircling the black hole. Matter close to the inside edge of the disc will eventually spiral into the black hole, but energy released by infalling matter and by frictional effects within the disc will heat it to very high temperatures, causing it to emit copious quantities of radiation. Because the supermassive black hole is so small compared with a galaxy, and most of the energy is radiated from the inner parts of the swirling accretion disc, this model neatly accounts for the very compact nature of the energy sources in AGNs. Rotating hot spots and fluctuating inflows of material allow the power output to fluctuate rapidly (Fig. 8.13).

By a process that is still not fully understood, but may involve twisted magnetic fields generated in the neighbourhood of the underlying black hole, the central 'engine' accelerates streams of charged particles to extremely high speeds – large fractions of the speed of light. The inner part of the accretion disc, together with the surrounding gas and magnetic fields, form a nozzle that confines the outward flow of particles into narrow streams that shoot out perpendicular to the disc, thereby creating the jets that are observed in many radio galaxies and quasars.

If we are looking straight down, or very close to, the axis of the jet, the dazzling high-energy beam dominates our view. In those circumstances, we will see a compact, violently

Figure 8.13 Artist's impression of the dusty torus around a supermassive black hole at the centre of an active galaxy, and the jets of energetic particles that shoot out perpendicular to it. Image credit: ESA/NASA, the AVO project and Paulo Padovani.

variable blazar. Looking at a moderate angle to the jet, we will see a relatively unobscured compact central energy source – a quasar – inside the torus at the heart of the AGN. From a viewpoint closer to the plane of the torus, the central engine will be hidden and we will see instead the jets and lobes of a radio galaxy. A similar picture applies to the less energetic Seyfert galaxies.

The accreting supermassive black hole model seems to offer a neat and convincing explanation for the great majority of active galactic nuclei as well as, on a much more modest scale, accounting for the compact energy sources in the nuclei of more conventional galaxies such as the Andromeda galaxy and the Milky Way.

9 The evolving Universe

In 1912, American astronomer Vesto Melvin Slipher, of the Flagstaff Observatory in Arizona, embarked on a programme to measure Doppler shifts (*see* Chapter 11) in the spectra of galaxies (which at that time were known as 'extragalactic nebulae'). By the end of 1923, he had managed to measure the wavelengths of lines in the spectra of 41 galaxies and had shown that all but five of them exhibited **redshifts** (the measured wavelengths of their spectral lines were longer than would be the case if they were at rest relative to us), which indicated that they were receding from us at speeds of up to 1800 kilometres per second.

Throughout the 1920s, Edwin Hubble pressed on with his programme of measuring galaxy distances (*see* Chapter 8). By combining his own data on galaxy distances with Slipher's measurements of their Doppler shifts, he found that the redshifts in the spectra of galaxies are proportional to their distances (the greater the distance, the higher the redshift), a result which he set out in an epoch-making paper published in 1929. Hubble's relationship – which has since come to be known as the **Hubble law** – implied that all galaxies and clusters of galaxies (apart from members of the Local Group, which are bound together by their mutual gravitational attractions) are receding from us with speeds that are proportional to their distances; the greater the distance, the higher the speed. Hubble's results implied that the universe is expanding (Fig.9.1).

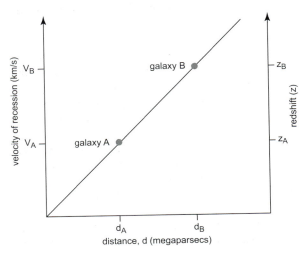

Figure 9.1 The Hubble law. The redshift in a galaxy's spectrum (z) and its speed of recession (V) are proportional to its distance. The red-shift and velocity of galaxy B, which is twice as far away as galaxy A, are twice as great as those of galaxy A.

The Hubble Law can be expressed as: velocity of recession (V) = distance (D) multiplied by a constant (H), which gives the numerical relationship between speed and distance. This constant is known as the **Hubble constant** (or more properly, the Hubble parameter). The value of the Hubble constant is usually expressed in units of kilometres per second per megaparsec (km/s/Mpc), where a **megaparsec** is a million parsecs and a parsec (*see* Chapter 6) is equivalent to 3.26 light years. Determining the value of the Hubble constant has been one of the most challenging and demanding tasks that astronomers have undertaken, and it is only in the past couple of decades that its value has been pinned down with real confidence. Currently, its value is taken to be about 68 km/s/Mpc. This implies that a galaxy at a distance of 1 megaparsec (3,260,000 light years) is receding at a speed of 68 km/s; a galaxy at ten times that distance (10 Mpc) is receding ten times faster (680 km/s), and so on.

Big Bang and the age of the Universe

If the galaxies are getting further apart now, they must have been closer together in the past, so if we look far enough back in time, they must at some stage have been exceedingly closely packed together. This observation provides the basis for the widely held view that the universe originated a finite time ago by erupting from an exceeding compressed state (perhaps from a singularity – a state of infinite compression) in an event that has come to be known as the **Big Bang**. The value of the Hubble constant gives a clue to the expansion time, or 'age', of the universe. If the galaxies have been receding at constant speeds since the Big Bang, the time taken by any particular galaxy, travelling at velocity V, to recede to its present distance (D) is just distance divided by velocity (D/V). A galaxy that is ten times further away from us is travelling ten times faster, and so will have taken exactly the same amount of time to reach its present distance. The expansion time of the universe calculated on the assumption that the galaxies have been flying apart at constant speeds is known as the **Hubble time**, and, for a Hubble constant of 68 km/s/Mpc, is just over 14 billion years.

At first glance the observed recession of the galaxies might seem to suggest that we are at the centre of the universe, and that everything else is rushing away from us in particular. This is an illusion. All the evidence suggests that each galaxy (or cluster of galaxies) is receding from every other one, that the entire universe is expanding, and that there is no unique, definable centre to this expansion. A useful analogy is to represent the whole of space by the surface of a balloon, and the galaxies by spots stuck to that surface (ignore the inside of the balloon and the world beyond its surface; think only of its skin). If the balloon is inflated to twice its initial size, the pattern of spots on its surface will remain the same, but the separation between each spot will be twice as great as it was before. Viewed from one particular spot (A), if another spot (C) were initially twice as far away as a nearer spot (B), it will still be twice as far away. From the point of view of spot A, the distance travelled by C, due to the expansion of the balloon, will be twice as great as the distance travelled by B; the observer will conclude that all the other spots are receding with speeds that are proportional to their distances and will arrive at the Hubble law. But no one spot can claim to be the centre of this expansion – each spot is receding from every

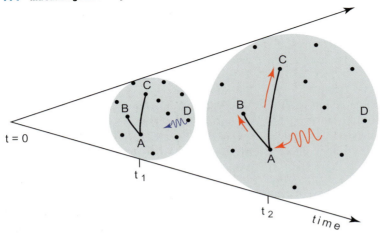

Figure 9.2 The balloon analogy. If the size of the balloon doubles between times t1 and t2, the distances between the spots (which represent galaxies) become twice as great. Seen from galaxy A, as indicated by the arrows, the increase in C's distance is twice as great as the increase in B's distance. The wavelength of light (blue) that was emitted by galaxy D at t1 has also doubled (red) by the time it arrives at galaxy A.

other one as the space between them (the surface of the balloon) stretches (Fig. 9.2).

Although the comparison should not be taken too far, the balloon analogy also provides another way of looking at galaxy redshifts. Instead of thinking of them as being Doppler effects caused by galaxies rushing away from each other *through* space, cosmologists ascribe redshift to the stretching effect of the expansion of space. The **cosmological redshift** relates directly to the amount by which the universe has expanded in the time interval between the instant at which light was emitted from a remote galaxy and the time at which that light arrives here on Earth. For example, if the separation between galaxies has doubled in the time it has taken for a light wave to reach us from a particular galaxy (so that we are seeing that galaxy as it was when the universe was half its present 'size'), the expansion of space will have stretched that wave to twice its previous wavelength. Consequently, the wavelength of each line in that galaxy's spectrum will be twice as great as would be the case if the galaxies were stationary and not

rushing apart. The observed redshift, therefore, is directly related to the amount by which the universe has expanded.

From the Big Bang to the present

The Big Bang was not an ordinary explosion in which matter was scattered forth from a particular point in a pre-existing empty space. Rather, space, time and matter all originated with the Big Bang and 'before' that event there was no space, time or matter in the sense in which we use those words today. Precisely how the universe came into existence in that first instant remains beyond the scope of present-day physics, but theoreticians believe that its history after the first microscopic fraction of a second can be described in terms of the known laws of physics. In outline, the story goes something like this:

Very soon (about 10^{-35} seconds) after the initial event the universe underwent a brief, but dramatic, phase of accelerating expansion – called **inflation** – which probably lasted for no more than 10^{-32} seconds, but during which the size of the universe grew by a colossal factor (of at least 10^{27}, and possibly much more). Thereafter, the universe reverted to expanding at a decelerating rate. At the end of the short-lived inflationary era, the universe was still exceedingly hot, dense, and filled with intense high-energy radiation. In accordance with Einstein's relationship between mass and energy ($E=mc^2$), energetic radiation could transform into pairs of particles and **antiparticles** (particles with 'mirror image' properties); conversely when a particle collides with its antiparticle, the two are annihilated and transformed back into photons of radiant energy. In order to make particle–antiparticle pairs with a particular mass, the energies of the photons – which depended on the temperature of the mix of matter and radiation that permeated the universe – had to exceed a particular threshold.

As the universe continued to expand and cool down, photon energies quickly dropped below the thresholds at which the more massive particles could be formed. **Baryons** (particles such as protons, neutrons and the **quarks** of which they are constructed) ceased to be created about a millionth of a second after the beginning of time, when the temperature dropped below about 10^{13} K. The overwhelming majority of baryons and **antibaryons** quickly collided and annihilated each other, turning into photons of radiation. Had there been an exact equality between the number of particles and antiparticles, virtually all the matter in the universe would have been annihilated, and there would be no galaxies, stars, planets or people in the universe today. By comparing the number of photons to the number of baryons in the present day universe, cosmologists have deduced that when the mutual annihilation was taking place, there

were about a billion and one baryons for every billion antibaryons. The entire matter content of the universe today is the one in a billion residue of the orgy of self destruction that took place about a millionth of a second after the beginning of time.

A few seconds later, when the temperature had dropped to around 5×10^9 K, electrons and positrons (the **positron** is the antiparticle of the electron) ceased to be formed and they in turn mutually annihilated, leaving a one in a billion residue of electrons in a sea of photons, protons, neutrons and more exotic particles. A few seconds later, when the temperature had dropped to about 10^9 K, collisions between protons and neutrons resulted in fusion reactions that began to create nuclei of helium, together with small quantities of deuterium ('heavy hydrogen'; a deuterium nucleus consists of one proton and one neutron), lithium, beryllium and boron. During the next few hundred seconds, about a quarter of the total mass of ordinary (baryonic) matter was transformed into helium, leaving about three quarters in the form of hydrogen.

Thereafter, the universe consisted of an expanding, cooling mix of matter and radiation, which became rapidly more dilute. Because photons could travel hardly any distance before colliding with fast-moving electrons or atomic nuclei, space was at that time opaque to electromagnetic radiation. That situation changed dramatically about 380,000 years after the Big Bang, by which time the temperature everywhere had dropped to about 3000 K. For the first time in the history of the universe, atomic nuclei were able to capture, and hold onto, electrons to make complete atoms. The electrons, which were mainly responsible for scattering photons and making the universe opaque to light, were mopped up quickly so that, in a short time on the cosmic scale, space changed from being opaque to transparent. Photons were then able to spread freely through the expanding volume of space. Since then, the expansion of space has cooled and diluted this primordial radiation and has stretched its wavelengths by a factor of more than a thousand, turning it from visible and infrared radiation (which it was at the time when matter and radiation separated out) to microwaves, with wavelengths of about a millimetre, which corresponds to a temperature of 2.7 K – just under three degrees above Absolute Zero.

If this version of events is correct, the universe should be permeated with a faint background of microwave radiation spread uniformly across the sky. The serendipitous detection of this **cosmic microwave background radiation** (the CMBR) in 1964 by Arno Penzias and Robert Wilson (who were working on communication systems for Earth satellites at the time) was a crucial breakthrough in the validation of the hot Big Bang theory. Although the CMBR is remarkably smooth and isotropic (it looks the same, apart from very minor differences, in all directions), it does contain a pattern of marginally warmer and cooler patches (called **temperature fluctuations**) which is the fingerprint of underlying marginally denser and less dense regions (**density fluctuations**) in the primordial mix of matter and radiation at the time when the background radiation was released, some 380,000 years after the Big Bang. These were the seeds from which all kinds of present-day structures – from small galaxies to giant superclusters – subsequently evolved (Fig. 9.3).

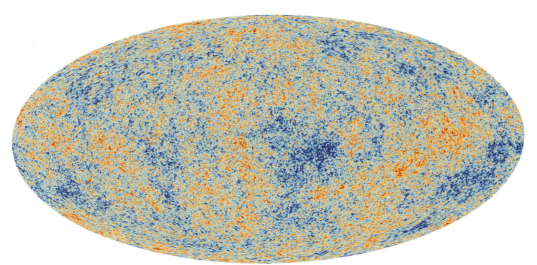

Figure 9.3 This all-sky map of the cosmic microwave background radiation, from data acquired by the Planck spacecraft, shows tiny temperature fluctuations (cooler and warmer spots) that correspond to regions of slightly different density from which galaxies and clusters eventually formed. Image credit: ESA and the Planck Collaboration.

In recent years, detailed investigations of the CMBR have revealed that the first stars, quasars and other luminous objects had formed and begun to shine just a few hundred million years after the Big Bang. Furthermore, in March 2014, a team called the BICEP2 Collaboration announced the discovery of an additional, exceedingly subtle, pattern imprinted on the CMBR by **gravitational waves** (wave-like gravitational disturbances which stretch and squeeze space) that rippled through space during, and after, the inflationary era. In addition to providing direct evidence for the existence of gravitational waves (which were predicted by Einstein), this discovery, if confirmed, provides strong evidence that inflation did indeed occur in the very early universe, a microscopic fraction of a second after the beginning of time.

Studies of the CMBR, together with improved determinations of the Hubble constant and measurements of changes in the rate of expansion over cosmic history, have pinned down the age of the Universe – the time since the Big Bang – at 13.8 billion years. All in all, the detailed analysis of the CMBR has provided a treasure trove of information about: conditions in the early universe, the relative proportions of its various constituents, its overall mean density, the way in which matter is distributed, the way in which galaxies and large-scale structures formed and evolved over time, the extent to which space is curved by matter, and has furnished important clues to what may happen to the Universe in the distant future.

Inventory of the Universe

Studies of the information encoded in the CMBR, together with a wide range of other observations, have led to the somewhat surprising conclusion that the universe is composed of three principal ingredients: ordinary **baryonic matter** (the stuff of which atoms, stars, planets and people are made), non-baryonic dark matter (matter composed of particles that are completely different in nature from the protons and neutrons that make up atoms) and an extra ingredient, called **dark energy**, the nature of which is completely unknown. According to data obtained by the Planck spacecraft, and released in 2013, the relative proportions of these ingredients are: baryonic matter – 4.9%; dark matter – 26.8%; and dark energy – 68.3%. To our considerable consternation, we have to admit that we do not know the nature of about 95% of the stuff of which the universe is composed.

We know that large quantities of dark matter exist within and around galaxies and galaxy clusters (*see* chapter 8) and there can be little doubt that dark matter has played an important role in the formation and evolution of galaxies, clusters and large-scale structures in the universe. Although we do not yet know the precise nature of dark matter, the most popular idea is that it consists of elementary particles called **WIMP**s (an acronym for 'weakly-interacting massive particles') which are heavier than protons or neutrons (possibly about 50–100 times the mass of the proton) but which do not interact with each other, or with other particles, through the agency of the **strong nuclear force**, which binds together the quarks that make up baryons, and which holds protons and neutrons together within atomic nuclei; nor do they interact via the **electromagnetic force**, which controls the absorption and emission of light. But exceedingly rarely, a WIMP may collide with an atomic nucleus, causing it to recoil. For more than two decades now, various research teams have been attempting to detect these kinds of events. To shield the detectors from extraneous influences (such as impacts by cosmic rays) that would produce spurious results, these experiments are located deep underground. As yet, no clear-cut detection of a WIMP has been achieved. Another possibility is that dark matter particles may eventually be created and detected inside powerful particle colliding machines such as the Large Hadron Collider at CERN in Geneva. Alternatively, WIMPs may reveal their presence indirectly, through the eventual detection of particles and radiation produced when dark matter particles decay or collide with each other.

The formation and evolution of galaxies and clusters

The seeds for galaxy formation were clumps of matter that were marginally denser than their surroundings. Because of the pull of its own gravity, a clump that was denser than average would expand progressively more slowly than its surroundings; eventually it would cease to expand, and would collapse on itself to form a self-contained structure. The most favoured hypothesis is that small **protogalaxies** (precursors of galaxies) formed first (Fig. 9.4), and then grew through a succession of collisions, mergers and galactic cannibalism (where one galaxy strips material from, and eventually absorbs, a smaller one). This process is still going on in the universe today.

Figure 9.4 This image shows three primitive galaxies embedded within a giant primordial bubble of gas. Seen as they were when the universe was only 800 million years old (6% of its present age), these infant galaxies appear poised to merge into a single massive galaxy. Image credit: NASA, ESA, ESO, NRAO, NAOJ, JAO.

Acknowledgements: NASA, ESA, ESO, NRAO, NOAJ, JAO, M. Ouchi (University of Tokyo), R. Ellis (California Institute of Technology), Y. Ono (University of Tokyo), K. Nakanishi (The Graduate University of Advanced Studies (SOKENDAI) and Joint ALMA Observatory), K. Kohno and R. Momose (University of Tokyo), Y. Kurono (Joint ALMA Observatory), M. Ashby (Harvard-Smithsonian Center for Astrophysics), K. Shimasaku (University of Tokyo), S. Willner and G. Fazio (Harvard-Smithsonian Center for Astrophysics), Y. Tamura (University of Tokyo), and D. Iono (National Astronomical Observatory of Japan).

Dark matter is believed to have played a crucial role in this story. Because ordinary matter particles and photons interact strongly with each other, clumps of ordinary matter could not begin to pull themselves together under the action of gravity until after the close coupling between photons and baryons was broken – some 380,000 years after the Big Bang. On the other hand, dark matter particles, because they do not interact with photons, could readily begin to clump together much earlier; consequently dark matter clumps are likely to have acted as the 'gravitational wells' into which ordinary matter subsequently fell to create the precursors of galaxies and clusters. This allowed the growth of structure to proceed much more rapidly than would have been the case with ordinary matter alone.

The evolution of structure in the universe has been modelled by numerical simulations in supercomputers. These indicate that dark matter aggregated together to form a network of clumps and threads, into and along which ordinary matter flowed and accumulated to form galaxies, clusters and the panoply of superclusters, sheets and strings of luminous matter that we see in the universe today (Fig. 9.5).

The future of the universe

Will the universe continue to expand forever? If matter (both luminous and dark) were

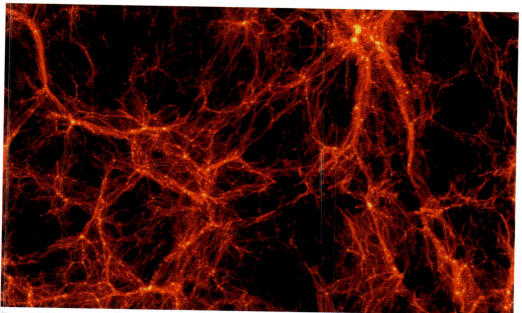

Figure 9.5 This computer simulation shows the predicted distribution of dark matter in the universe. Galaxies form where denser concentrations arise within the filaments of dark matter. The region shown here is a billion light-years across. Image credit: Prof. Dr. Ben Moore, Centre for Theoretical Astrophysics and Cosmology, University of Zurich, Switzerland.

the only constituent of the universe, then whether or not the universe will continue to expand forever would depend on the mean density of matter averaged over the whole of space. If the mean density of the universe is less than a particular value, which is called the **critical density** and is equivalent to about five hydrogen atoms per cubic metre of space, then gravity will not be strong enough to halt the ongoing expansion. Although the rate of expansion would slow down over time, it would never drop to zero, and the expansion would go on forever. If, on the other hand, the overall mean density were greater than the critical density, then the expansion would slow to a halt at a finite time in the future and, thereafter, gravity would cause all the galaxies (or what is left of them by then) to begin to fall together, slowly at first, but ever faster until at some future date, all of the matter in the universe would merge together and the universe would end in a **Big Crunch** – a state of infinite compression and extreme temperature, rather like the Big Bang in reverse.

A universe in which the mean density is precisely equal to the critical density would just, but only just, be able to expand forever, the speeds at which the galaxies were separating dropping ever closer to, but never quite reaching, zero. A universe that sits on the fence like this, poised between the open, ever-expanding option and the closed, ultimately collapsing, one is called a **flat universe** because the net curvature of space caused by the matter and various forms of energy that it contains is zero; space in such a universe is flat in the sense that rays of light (apart from local deviations caused by individual clumps of matter) will travel through space in straight lines.

A wide range of different kinds of observational evidence have shown that the overall mean density of matter and energy in the universe is indistinguishably close to the critical density, and that space is flat. However, ordinary (baryonic) matter accounts for just under five percent of the critical density, and dark matter a further 27% so that in total, matter provides only about 32% of the critical density. The rest must consist of something else – an extra ingredient that has come to be known as dark energy.

Compelling evidence for the existence of dark energy has been provided by observations which show that, far from slowing down, the expansion of the universe appears instead to be speeding up. The first evidence for this extraordinary assertion came in 1998 from two teams of astronomers (The High-redshift Supernova Search and the Supernova Cosmology Project) who – independently – were measuring the distances of remote galaxies by identifying Type Ia supernovae (*see* Chapter 7) embedded within them. Type Ia supernovae are a particularly good 'standard candle' because, within a small amount of variation, they all appear to attain the same peak luminosity. Astronomers can measure their distances, and the distance of the galaxies within which they are embedded, by comparing their observed apparent brightnesses with assumed values for their inherent peak luminosities. What the two teams found was that distant supernovae in galaxies with high redshifts were systematically fainter (and hence further away) than would be the case if the universe were expanding at a steady rate, or if the expansion were slowing down. They arrived at the startling conclusion that over the past several billion years, the expansion of the

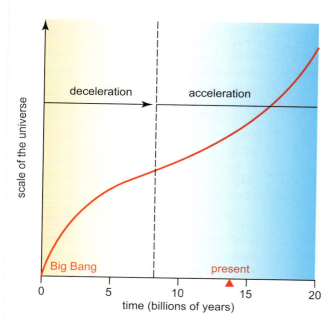

universe has been accelerating. Some extra force of repulsion, opposed to gravity, must be responsible for this behaviour. This extra ingredient, a form of energy that permeates the universe and exerts a repulsive influence, is dark energy.

Over the past decade and a half, further, more precise, studies of supernovae and other indicators of galaxy distances, which probe the history of the universe further and further back in time, have confirmed that the universe appears to have switched from decelerating under the action of gravity to accelerating under the influence of dark energy some five to six billion years ago (Fig. 9.6). When the universe was younger and matter was more densely concentrated than it is now, gravity was dominant; but as the universe expanded further, and matter became more thinly spread, its attractive influence eventually

became less powerful than the repulsive effect of dark energy, and the expansion began to accelerate.

The nature of dark energy is completely unknown. One possibility is that it is related to a quantity called the **cosmological constant**, which Einstein invoked in 1917 to account for why the universe was – as was believed at that time – static, rather than expanding or contracting. This extra ingredient was, in effect, a form of cosmic repulsion which, Einstein thought, would have counterbalanced the attractive effects of gravity and prevented the stars, and what we now call galaxies, from falling together. After the discovery of the expansion of the universe in the late 1920s, Einstein rejected the cosmological constant, and is said to have remarked that it was 'his greatest blunder'. Nowadays, the cosmological constant, in more modern guise (where it is regarded as a type of '**vacuum energy**' – a form of energy that permeates the universe, that is constant and unchanging, and has a repulsive influence), is seen as one of the front-running candidates for what dark energy may be. But there are many alternative hypotheses. Depending on the nature of dark energy, and whether it remains constant or changes over time, several very different futures are possible.

If the repulsive effect of dark energy remains constant, or evolves in such a way that it becomes weaker but never declines to zero, the accelerating expansion will carry on forever: galaxies will continue to separate ever faster, their constituent stars will pass through their life cycles and fade to obscurity, and the universe will evolve to a cold, dark future, known, somewhat graphically, as the **Big Chill**. By contrast, if dark energy's repulsive influence grows with time, the expansion will accelerate so violently that clusters and galaxies, then stars and planets, and then atoms, elementary particles and the fabric of space itself will be torn apart in a catastrophic **Big Rip** at some time in the future. Or again, perhaps dark energy's repulsive influence will eventually decline to zero and transform into an attractive one, which acts in concert with gravity. In that case, the expansion would eventually cease, and then – with gravity and dark energy acting together – the universe might, after all, collapse into a Big Crunch.

Determining the properties and nature of dark energy is one of the greatest challenges, if not the greatest challenge, facing modern physics. Until such time as the true nature and precise properties of dark energy can be pinned down, we cannot know which, if any, of these possible fates is the one that awaits our Universe. Indeed, it is possible that we may never know.

Table 9.1 Timeline of the universe.

Time	State of the universe / key events
0	Big Bang. Expansion of the universe begins from state of extreme (infinite?) density and temperature.
10^{-35} sec	Inflation begins. Universe undergoes very short-lived, but dramatic, period of accelerating expansion, which is called inflation.
10^{-32} sec	Inflation ends. Quarks form.
10^{-6} sec	Protons and neutrons form.
1 sec	Nuclear fusion (nucleosynthesis) begins.
3 min	As a result of nucleosynthesis, universe now contains hydrogen and helium nuclei in ratio 3:1 by mass, together with traces of deuterium, lithium, beryllium and boron.
3.8×10^5 yr	Hydrogen atoms form. Universe becomes transparent – cosmic background radiation spreads through expanding volume of space.
4×10^8 yr	Earliest stars and protogalaxies form.
10^9 yr	Milky Way and other substantial galaxies and structures form.
8×10^9 yr	Expansion of the universe begins to accelerate.
9.2×10^9 yr	Solar System forms (4.6×10^9 years ago).
13.8×10^9 yr	Now

10 Exoplanets and the conditions for life

During the past few years, astronomers have made dramatic advances in finding **exoplanets** (or extrasolar planets) – planets that revolve around other stars. To see them directly is exceedingly difficult, because planets are so much fainter (typically a billion times fainter) than the stars around which they revolve and, therefore, are hidden by the glare of their host stars. Only a few tens of exoplanets have been imaged. The overwhelming majority have been discovered either by measuring their gravitational influence on their host stars or by means of transits.

In the simple case where a star has a single planet, the planet and star will revolve round their common **centre of mass** – a point which lies on the line joining the centre of the planet to the centre of the star at a location that depends on their relative masses. For example, if the planet had one-thousandth of the star's mass (as is the case with Jupiter and the Sun), the centre of mass would lie one thousand times closer to the star than the planet. In principle, it ought to be possible to detect the very slight periodic wobble in the star's motion caused by the gravitational influence of the orbiting planet (the planet–star system would behave like an astrometric binary – *see* Chapter 6). It is exceedingly difficult to measure such tiny movements accurately and, as a result, only a tiny handful of planets have been detected in this way.

However, the small periodic Doppler shift in the star's spectrum caused by its orbital motion round the centre of mass is much easier to detect. When the star is on one side of the centre of mass, it will be moving away from us and its spectral lines will be red-shifted, and when it is on the other, it will be moving toward us, and its lines will be blue-shifted. The lines will shift to and fro relative to their mean wavelengths in a cyclic fashion with a period equal to the orbital period of the planet around the star (Fig. 10.1). The first planet to be discovered by this spectroscopic technique was the companion to a star in Pegasus, called 51 Pegasi, a star that is similar in nature to the Sun and which lies at a distance of about 48 light years. The planet, which is known as 51Pegb, is very different from the Earth. Its orbital period of 4.2 days indicates that it lies very close to its parent star – just 0.05 AU distant; consequently, its surface temperature is likely to be about 1300 K.

It is hard to obtain a precise value for the mass of a planet using this 'Doppler' technique because the angle at which the orbit is tilted relative to our line of sight is not usually known. If the orbit lies exactly in the line of sight, so that we are viewing it edge on, then measurements of the maximum Doppler shifts in its spectrum, which occur when the planet is moving directly towards or away from the Earth, will yield the true orbital velocity of the planet. If the orbital velocity and the orbital

A

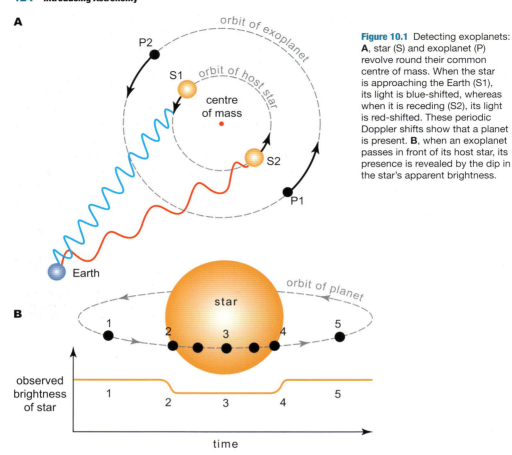

Figure 10.1 Detecting exoplanets: **A**, star (S) and exoplanet (P) revolve round their common centre of mass. When the star is approaching the Earth (S1), its light is blue-shifted, whereas when it is receding (S2), its light is red-shifted. These periodic Doppler shifts show that a planet is present. **B**, when an exoplanet passes in front of its host star, its presence is revealed by the dip in the star's apparent brightness.

period are both known, the mass of the planet can be calculated as a fraction of its host star's mass. If, however, the orbital plane of a planet were perpendicular to our line of sight (i.e., if we are looking directly down on the plane of its orbit), the planet would neither approach nor recede from us, and no Doppler shift would be detected. Usually, the plane of the orbit will lie at an unknown angle to the line of sight, and all that can be calculated from the period and the changing **radial velocity** (velocity directly towards or away from the Earth) deduced from the Doppler shifts is the *minimum* possible mass for the planet.

The mass of 51Pegb is estimated to be at least 0.47 Jupiter masses (i.e. at least 150 times that of the Earth). The discovery of a Jupiter-mass planet so close to its parent star came as a great surprise at the time, but since then a great many '**hot Jupiters**' (high-mass planets close to their host stars) have been found. This is not too surprising because it is much easier to detect high-mass planets close to their parent stars than lower mass ones, or high-mass ones much further away. The reason for this is that high-mass stars orbiting rapidly close to their parent stars have a much greater influence on the star's motion than slow-moving planets further away and, because they have very short periods, relatively short observation times are required in order to reveal their presence. Low-mass planets – comparable to the Earth – are much harder to detect, because their effects on their host stars are so much smaller.

Another technique that has been very productive is planetary **transits**. If a planet passes in front of its host star, it will blot out a small proportion of that star's light (Fig. 10.2). For example, if astronomers in a distant planetary system were to observe a transit of Jupiter across the face of the Sun, they would see the Sun's apparent brightness drop by about one percent (Jupiter has about a tenth of the Sun's diameter and its apparent area is about one percent of the area of the Sun's visible disc; therefore it would blot out about one percent of the Sun's light). If the Earth were to transit in the same way then, because the diameter of the Earth is slightly less than one-hundredth

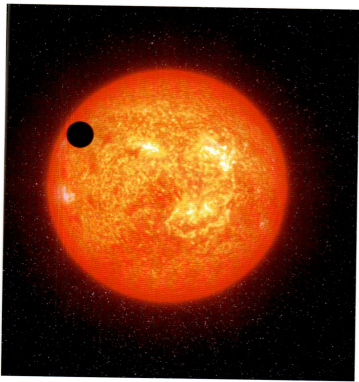

Figure 10.2 An artist's impression of the transiting exoplanet GJ 1214b passing in front of its faint red parent star. This planet, has a mass about six times that of the Earth. Image credit: ESO/L. Calçada.

of the Sun's diameter, the area it would blot out is less than one ten-thousandth of the visible disc of the Sun and the drop in brightness would be less than one-hundredth of one percent. The distant astronomers would see the one percent drop in brightness due to Jupiter once every 11.86 years (the orbital period of Jupiter) and, if they could detect it at all, the 0.01% dip caused by the Earth at intervals of one year.

The transit method has been used very successfully over the past few years by ground-based and space-borne instruments. For example, it was the chosen modus operandi for NASA's Kepler spacecraft, which – between 2009 and February 2014 – had detected 246 confirmed planets, together with more than 3600 possible detections that are awaiting confirmation.

The great majority of planets have been detected using either the radial velocity (Doppler) method or the transit technique. Although the chance of a planetary system being orientated at such an angle that a transit can be seen is small, one great advantage of the technique is that the diameter of the planet can be deduced from the light-curve of the transit (the graph of how the measured brightness of the star changes while the planet is crossing in front of it). When this information is combined with radial velocity data, which reveals the planet's mass (its orbital plane has to be close to edge on for a transit to occur), the mean density of the planet can be calculated. A low density would be consistent with a Jupiter or Neptune-like planet, whereas a high density would be consistent with a terrestrial type rocky planet. Furthermore, when a transiting planet is crossing the face of its host star, the star's light passes through its atmosphere; high precision spectroscopy can reveal absorption lines superimposed on the spectrum of the star by the gases present in the planet's atmosphere.

The first transiting planet to be detected (in 1999) was the hot Jupiter, HD 209458b. Subsequent spectroscopic observations showed that this planet has an atmosphere that contains hydrogen, oxygen, carbon, water vapour, sodium, methane, titanium oxide and carbon dioxide. Revolving round a G0 star similar to, but slightly larger and hotter than, the Sun in just 3.5 days at a distance of 0.045 AU, its surface temperature is believed to be about 1130 K, and there is clear evidence that its atmosphere is evaporating off into space (Fig. 10.3). The first Earth-sized exoplanet with a rocky composition to be identified was Kepler-78b, which is 1.2 times the size of the Earth and 1.7 times more massive; these figures imply that its density is the same as the Earth's, and indicate that it is composed mainly of rock and iron. Unfortunately, with an orbital period of 8.5 hours, its distance from the centre of its host star is less than 0.01 AU and its surface temperature must be more than 2200 K.

As of April 2014, the total number of confirmed exoplanets was in excess of 1700, these being distributed among more than 1100 planetary systems, 460 of which contain two or more planets. With a further 3800 candidates as yet unconfirmed, the total number of confirmed and 'possible' exoplanets found so far is well over 5000, with new discoveries being added rapidly. Although the greatest proportion of *confirmed* exoplanets are Jupiter mass bodies, of the 3538 candidate planets discovered by the Kepler mission up until November 2013, about 19% were Earth-size (less than 1.25 Earth radii), 30% were **super-Earths**

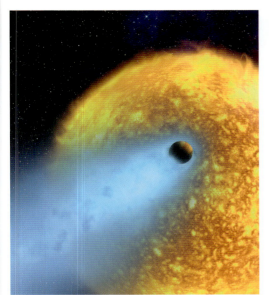

Figure 10.3 Artist's impression of exoplanet HD 209458b. This giant planet is so close to its Sun-like host star that its temperature is more than 1100 K, and its atmosphere (shown in blue) is evaporating off into space. Image credit: European Space Agency, Alfred Vidal-Madjar (Institut d'Astrophysique de Paris, CNRS, France), and NASA.

(1.25–2 Earth radii), 41% were Neptune-sized (2–6 Earth radii), 6% were Jupiter-sized (6–15 Earth radii) and 3% were super-Jupiters (larger than 15 Earth radii).

Extrapolating from these statistics, it seems likely that the Milky Way galaxy contains at least as many planets as stars (i.e. at least 200 billion planets) and that low-mass planets (Earths and super-Earths) are the most abundant. Our own Solar System has eight planets. Already one extrasolar system has been shown to possess at least seven planets, several contain at least six, and of the multi-planet systems confirmed so far about two-thirds of them contain three or more planets.

Conditions for life – in the Solar System and beyond

Although it is conceivable that there may be forms of life that are utterly different in nature, structure and composition to that with which we are familiar, when trying to discuss whether or not life exists elsewhere, we have to focus on the conditions necessary for life as we know it. Life on Earth is based on the ability of the element carbon, in combination with elements such as hydrogen, nitrogen and oxygen, to form the complex molecules, chains and self-replicating structures that have enabled living organisms to form, grow, evolve and pass on the coded genetic information that is needed to reproduce their own kind. Life requires an energy source (such as a star that provides radiant energy) and the presence of a solvent within which reactions that construct the requisite molecules and structures can take place efficiently. For life as we know it, water is that solvent.

The range of distances around a particular star within which, in principle, liquid water could exist on the surface of a planet, is called the **habitable zone** (Fig. 10.4). Whether or not a planet located within that zone actually has liquid water depends on several other factors, including atmospheric composition and pressure, and albedo (reflectivity); a highly reflective planet will absorb less starlight, and will therefore be cooler, than a planet of low reflectivity. Within the Solar System, the habitable zone – within which the Earth could be moved and still potentially have water somewhere on its surface – extends from about the orbit of Venus to around the orbit of Mars. Although conditions are self-evidently right on the Earth, Venus, with its powerful greenhouse effect, is much too hot and has no liquid

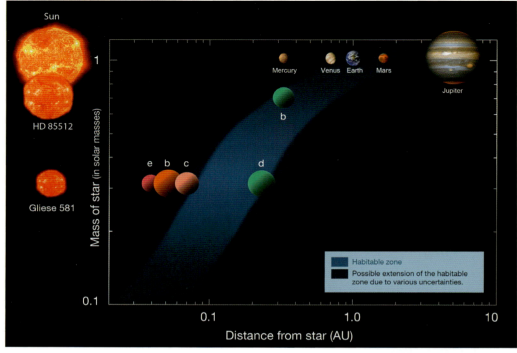

Figure 10.4 Habitable zones: This diagram shows the distances from their respective stars of the planets in the Solar System (top row), the HD 85512 system (middle) and the Gliese 581 system (bottom row). The habitable zone is indicated by the blue area. Image credit: ESO.

water, whereas Mars is too cold, and its atmospheric pressure too low, to allow liquid water to exist on its surface, although water ice is certainly present in its polar caps, and trapped within some of the surface rocks and soils (*see* Chapter 5).

Whereas Venus seems unlikely ever to have supported life of any kind, Mars may have done so in the remote past when conditions there were more favourable than they are now. Another possibility, so far as the Solar System is concerned, is that conditions suitable for life of some kind may exist beneath the surfaces of some of the ice-rich satellites of the giant planets – in particular, Jupiter's satellite Europa and Saturn's moon Enceladus. Both of these satellites lie well outside what normally would be considered the habitable zone for a sun-like star, which suggests that the concept of a habitable zone cannot be prescribed too strictly.

Planets suitable for life are most likely to exist around certain types of stars. The highly luminous O and B type stars are almost certainly unsuitable as hosts for life-bearing planets. Apart from radiating copious amounts of intense ultraviolet radiation – which is harmful to living organisms as we know them – these stars are very short-lived, their main-sequence lifetimes being barely long enough for planets to form, let alone to allow life to originate and develop on their surfaces. At the other end of the range, dim, cool M-type main-sequence stars (the most common type of star) have other problems. Although their long lifespans provide plenty of time for life to emerge and evolve, their very low luminosities imply that potentially life-bearing planets would have to lie very close to their host stars. The most suitable stars are of types G and K, which make up at least ten percent of the total number of stars in our galaxy; this suggests that there ought to be tens of billions of suitable host stars in our galaxy alone.

Hunting for 'Goldilocks planets'

Are there any '**Goldilocks planets**' out there, where the temperature is not too hot, not too cold, but 'just right' (like baby bear's porridge in the story of Goldilocks and the Three Bears)?

The current tally of confirmed exoplanets includes at least 10–15 planets with less than twice the Earth's diameter that lie within their host stars' habitable zones. Among the smallest habitable zone planets discovered by Kepler are two of the five planets that revolve around the star Kepler-62, a cooler, older star than the Sun. Of these planets, Kepler 62-f, which is about 40% larger than the Earth, lies well within the habitable zone, and Kepler-62e, which is 60% larger than the Earth, orbits at its inner edge. In June 2013, a team of astronomers from the European Southern Observatory found that the star Gliese 667C has no fewer than six planets, three of which lie within its habitable zone; all three are super-Earths – more massive than the Earth

but less massive than Neptune. However, at the time of writing, the only genuinely Earth-sized exoplanet that has been found within the habitable zone of its host star is Kepler 186f, the discovery of which was announced in April 2014. With a diameter almost identical to that of the Earth, this planet revolves around a cool M-type red dwarf star in a period of 130 days, and lies close to the outer edge of its habitable zone (Fig. 10.5). Its diameter has been established with confidence from transit data, but its mass and mean density have not yet been determined. Consequently, although it seems likely that Kepler 186f is a rocky world, its density and composition remain uncertain.

A statistical analysis of Kepler data has indicated that about 22% of sun-like stars (stars of spectral class G or K) in our galaxy are likely to host at least one Earth-sized planet (a planet with less than twice the Earth's radius) within their habitable zones. If this is correct, the nearest such star may well be no more than 12 light years distant from the Earth, and would be visible to the naked eye. When and if a true Earth-like rocky planet in the habitable zone of a Sun-like star has been discovered and confirmed, the next stage will be to try to build up as complete a picture of it as possible. If the spectral signature of its atmosphere reveals the presence of water, and its infrared emission indicates that its temperature is favourable, and especially if the atmosphere can be shown to contain significant amounts of oxygen, then such a planet would be a genuine candidate for the possible existence of life.

The search for extraterrestrial intelligence (SETI)

Given that planetary systems and solar-type stars are abundant, and that the appropriate chemical elements for life exist throughout

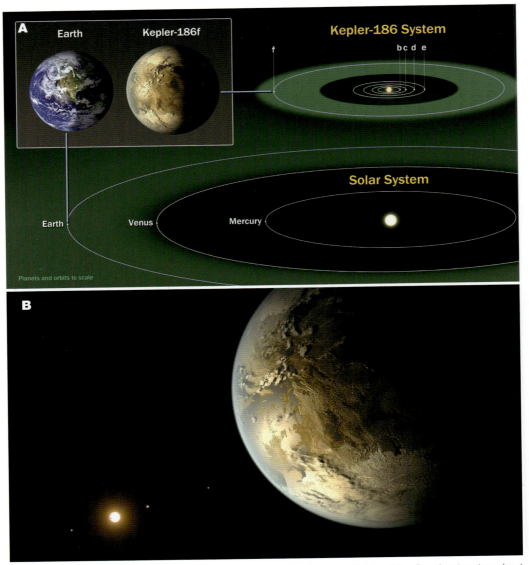

Figure 10.5 **A**, this diagram compares the planets of our inner Solar System to Kepler-186, a five planet system about 500 light-years from the Earth. The outermost planet – Kepler 186f – lies within this system's habitable zone. Image credit: NASA Ames/SETI Institute/JPL-Caltech. **B**, this artist's impression depicts Kepler-186f, which is almost identical in size to the Earth, and orbits near the outer edge of its host star's habitable zone in a period of 130 days. Image credit: NASA Ames/SETI Institute/JPL-Caltech.

the universe, it would seem highly unlikely (though not impossible) that life is a phenomenon unique to the Earth, and much more probable that it is widespread throughout the universe. If so, what is the likelihood of intelligent life having evolved on other worlds and of beings with the ability to communicate over interstellar distances existing elsewhere in our galaxy at this time? One could argue that the emergence of intelligence is an inevitable consequence of evolution, in which case there may well be significant numbers of advanced technological societies 'out there'. On the other hand, it may be that the chain of events that led to the evolution of intelligent life here on Earth was an exceptionally improbable one, in which case we may be alone in the galaxy.

Because the distances involved are so huge that sending probes to neighbouring planetary systems is out of the question for the foreseeable future, the only currently practicable way to search for evidence of the existence of **extraterrestrial intelligence** is to look for signals transmitted, deliberately or accidentally, by advanced civilizations. Such signals might be emitted in a variety of ways. For example, an alien civilization with unlimited resources might choose to announce its presence by means of an omnidirectional beacon, transmitting continuously or intermittently in all directions. Alternatively, an advanced civilization may be carrying out a programme of beaming signals towards suitable planetary systems, including, perhaps, our own, in hopes of getting a reply, or, if aware of the emergence of intelligent life on the Earth, may have chosen to beam signals specifically towards us. A further possibility is that we may pick up communications that are going on between advanced

civilizations where the line of sight between them happens to pass through the Solar System; in effect, we would be 'eavesdropping' on these 'messages'. Or, again, we may detect 'leakage radiation', the telltale output of a planet inhabited by a high-tech society (for example, television programmes).

At what frequencies should we be searching? Back in 1959, American physicists Giuseppe Cocconi and Philip Morrison suggested that, because neutral hydrogen emits radiation at a wavelength of 21.1 cm (a frequency of 1420 MHz), and because this would surely be known to any civilization that had developed radio astronomy, that would be the obvious frequency at which they would choose to transmit. The following year, Frank Drake undertook the first limited search at this frequency using a 26-metre radio dish at Greenbank, Virginia, to monitor two nearby sun-like stars – Tau Ceti and Epsilon Eridani. Although this two-month project – known as Project Ozma – detected no extraterrestrial signals, it acted as a spur to other groups, in various nations, who have carried out much more extensive searches in the subsequent decades.

The major present-day projects do not concentrate on the 1420 MHz hydrogen line itself; because hydrogen clouds in space radiate at that frequency, that frequency is too noisy. Particular interest is focused on the waveband between 1420 MHz (the hydrogen line) and 1660 MHz (the frequency at which the hydroxyl molecule – OH – radiates). In order to have any chance of picking up a signal, which is likely to be extremely weak, a receiver has to be tuned to precisely the right frequency. Modern searches use devices called multichannel spectral analysers, which can search millions, or hundreds of millions, of narrow frequency bands virtually simultaneously. Despite a small number of false alarms, so far no positively identifiable signal of alien origin has been recorded by any of the searches, although one signal, detected by the Ohio State University team in 1977, remains a tantalizing and intriguing mystery. The short-lived signal, which was about thirty times stronger than the natural background at 1420 MHz, tracked across the sky at the same rate as the stars, was intermittent, then terminated as if it had been switched off. Known as the 'Wow! Signal', because of a remark made at the time by team leader Jerry Ehman, it was never seen again.

Although, for good reasons, most searches have concentrated on looking for coded signals within the microwave region of the spectrum, evidence for the existence of ETI may instead be lurking in other parts of the electromagnetic spectrum. For example, because they operate at much higher frequencies, optical or infrared lasers have the capacity to carry much more information. They can be beamed from conventional telescopes, and can be restricted to extremely narrow wavebands so that if you happen to hit on just the right one, and if it were pointing precisely in our direction, the laser signal from a distant planet could – at that precise frequency – be brighter even than its parent star.

Searching for signs of extraterrestrial intelligence is rather like looking for a needle in a haystack without even knowing if the needle is there. It is not surprising, therefore, that tens or hundreds of thousands of hours of observing time have failed to come up – so far – with a clear-cut detection. Alien signals may not be there at all, or highly advanced civilizations may communicate by means of which we are as yet completely unaware. If the search eventually is successful, it will prove that alien intelligences exist – and that would be the most profound of scientific discoveries. But an unsuccessful search can never prove conclusively that we are alone.

Being proactive

If we pick up a message, should we reply, or would it be better to keep quiet? Or should we be proactive, and send out signals advertizing our presence? We have already done so, deliberately and inadvertently. For example, since the beginning of large-scale television broadcasting in the nineteen-fifties, evidence of our presence has already spread out (as exceedingly feeble signals) to a range of more than 60 light years. And there have been a number of deliberate attempts to send messages to other worlds. The first, in 1974, was transmitted from the 300-metre diameter fixed radio dish at Arecibo, Puerto Rico, towards the globular star cluster M13 in Hercules, which contains well over 100,000 stars. The three-minute transmission consisted of just 1679 bits of data, which, in a very simple mathematical way, could be decoded to reveal a basic image containing information about the planetary system from which it was sent, and the beings that sent it. Because M13 is about 24,000 light years away, it will take 24,000 for the signal to get there, and any reply would take a further 24,000 years; so we do not expect to get a response any time soon!

Among more recent terrestrial transmissions was a 'Message from Earth' broadcast in October 2008 towards the planet Gliese 581c, some 30 light years distant. This particular transmission, which was broadcast from a

Figure 10.6 Voyager's Golden Record. Each of the Voyager spacecraft (lower right), launched in 1977, carries a record with images and sounds of Earth. The cover of the record (upper right) carries instructions on how to play it and shows the location of our Sun. Image credit: NASA/JPL-Caltech.

70-metre radio dish in the Ukraine, contained 501 individual messages selected through a competition held on the social networking site, Bebo! In more passive fashion, the two Pioneer spacecraft (Pioneer 10 and 11) and the two Voyager spacecraft, all of which are heading out of the Solar System never to return, carry a plaque (in the case of the Pioneers) and a disc (plus needle with which to play it) in the case of the Voyagers, just in case an alien civilization should come across one of them one day (Fig. 10.6).

11 Tools of the trade

Astronomy is an observational science. Apart from studying meteorites that have fallen to the Earth, or samples of material from the Solar System brought back to the Earth, or investigated directly *in situ* by spacecraft, astronomers cannot carry out direct experiments on the objects they are studying – they cannot, for example, compress a star to see what happens. All that they can do is to detect, measure and analyse light and other forms of radiation or energetic particles that come our way from distant sources such as planets, stars and galaxies.

Electromagnetic radiation and the electromagnetic spectrum

Light is a form of **electromagnetic radiation**, an electric and magnetic disturbance that propagates through space at a constant speed – known as the **speed of light** and denoted by the symbol, c – which, in a vacuum, is just under 300,000 km/s. It can be regarded as a wave motion, similar in some respects to a wave on water. The distance between two successive wave crests is the **wavelength**, the 'height' of each wave crest above, or below the mean level, is the **amplitude**, and the number of wave crests per second that pass a particular point, or observer, is the **frequency** – the higher the frequency (the more wave crests per second), the shorter the wavelength. Light also behaves in some respects like a stream of tiny particles, called **photons**, which travel at the speed of light. Each photon carries a quantity of energy – the shorter the wavelength (and the higher the frequency) the greater the energy of the photon.

Visible light spans a range of wavelengths from just under 400 nanometres (a **nanometre**, symbol nm, is one billionth of a metre) to about 700 nm, and our eyes respond to different wavelengths by seeing different colours: from the longest visible wavelengths to the shortest, the principal colours that we perceive are red, orange, yellow, green, blue, indigo and violet – the colours of the rainbow. Wavelengths that are longer than visible are called **infrared**, whereas those which are shorter than visible are called **ultraviolet**. The complete range of wavelengths from the very shortest to the extreme longest is known as the **electromagnetic spectrum**, and is divided by convention into the following broad subdivisions: gamma ray (the shortest wavelength and highest energy form of electromagnetic radiation), x-ray, ultraviolet, visible, infrared, microwave and radio (the longest, and least energetic). The electromagnetic spectrum is depicted in Figure 11.1.

Refraction, dispersion and the spectroscope

When a ray of light passes from a less dense medium (such as air) into a denser one (e.g. glass) it is bent, or 'refracted'. The reason for this is that light travels more slowly through glass than it does through air or a vacuum.

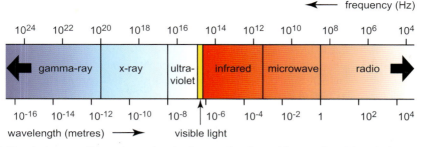

Figure 11.1 The electromagnetic spectrum, showing the wavelengths and frequencies of the principal regions into which it is divided. Visible light occupies only a tiny portion of the full range of wavelengths.

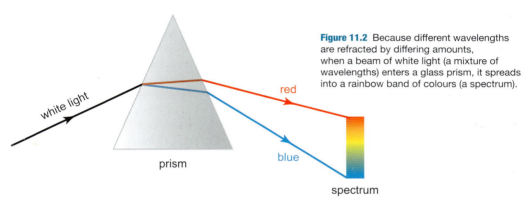

Figure 11.2 Because different wavelengths are refracted by differing amounts, when a beam of white light (a mixture of wavelengths) enters a glass prism, it spreads into a rainbow band of colours (a spectrum).

Different wavelengths of light travel in glass at different speeds, so they are refracted by differing amounts: the shorter the wavelength, the greater the **refraction** (e.g. blue light is refracted more than red). If a beam of white light – a mixture of all wavelengths and colours – passes through a glass prism, the different amounts of refraction experienced by the different wavelengths cause them to diverge from each other and spread out into a rainbow band of colours from red (least refraction) to violet (greatest refraction) (Fig. 11.2). This property of light (called **dispersion**) forms the basis for the **spectroscope** – a device that splits light into its constituent wavelengths

and spreads it into a band of colours, which is called a **spectrum**. A similar effect occurs when light passes through, or reflects from the surface of, a **diffraction grating** – a closely spaced grid of parallel lines or grooves that has been imprinted on the surface of an optical material. In a working spectroscope, light passes through a narrow slit before entering the prism (or being split up by a diffraction grating), and a system of lenses or mirrors produces a focused image of the resulting spectrum that may be viewed visually through an eyepiece (in a spectroscope), or imaged photographically or electronically (in a **spectrograph**).

As we saw in Chapters 6 and 7, the spectrum of a typical star consists of a rainbow band of colours (a continuous spectrum) upon which is imprinted sets of dark lines (absorption lines) which are characteristic of the various chemical elements of which it is composed, whereas a glowing gas cloud (an emission nebula) emits light at certain particular wavelengths only (an emission line spectrum). The study of spectra and spectral lines is the key to understanding the chemical and physical nature of all kinds of objects in the universe.

The Doppler effect

If a source of light is receding, each successive wave crest is emitted from a progressively greater distance, has to travel further than its predecessor, and therefore arrives later than it would have done if the source of light had been stationary. Fewer wave crests per second arrive at the observer than were emitted, per second, from the source. The frequency is reduced and the observed wavelength increased. In effect, light waves are stretched when the source is receding. Conversely, if the source is approaching the observer, the waves are compressed, the frequency is increased and the wavelength shortened. A similar effect occurs with sound waves, the pitch of an approaching source of sound being higher than the pitch of a receding one. This phenomenon is called the **Doppler effect**.

The Doppler effect changes the observed wavelengths of spectral lines, too. If a star or galaxy is receding, its pattern of lines is shifted to longer wavelengths; because red corresponds to the long-wave end of the visible spectrum, this phenomenon is called a **redshift**. If the light source is approaching, its pattern of lines appears at shorter wavelength

(this is a **blueshift**). The amount by which the wavelength of a line is changed, compared to the wavelength it would have if the source of light were stationary, is proportional to the speed at which the source is receding or approaching. By comparing the observed wavelengths of spectral lines with the wavelengths they would have if the source were stationary, astronomers can measure the radial velocities (speeds of approach or recession) of stars and galaxies.

Telescopes of various kinds

When parallel rays of light coming from a very remote object pass through a glass lens that has curved surfaces, the shape of the lens ensures that a ray that enters the centre of the lens perpendicular to its front surface carries on in a straight line, whereas all the other rays are refracted so as to converge to a point, called the **focus**, or **focal point** (Fig. 11.3). If the rays are coming from a source of finite angular (apparent) size, such as the Moon, they will form an image of finite size. For a simple lens (convex on both surfaces), the resulting image will be inverted (upside down). The distance between the centre of a lens and its focus is called the **focal length**, and the clear diameter of the lens is called the **aperture**. The longer the focal length of the lens, the larger (but fainter) the resulting image.

A **refracting telescope**, or **refractor**, uses a large lens of long focal length, called the **object glass**, or **objective**, to collect light and form an image. That image may then be magnified, and viewed by eye, with the aid of a second lens – the **eyepiece** – which has a short focal length (Fig 11.4). The **magnification**, or **magnifying power**, of the telescope (the ratio of the apparent size of an object when viewed

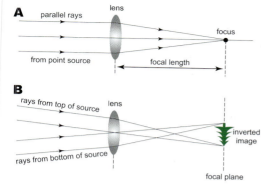

A parallel rays | lens | focus | focal length | from point source

B rays from top of source | lens | inverted image | rays from bottom of source | focal plane

Figure 11.3 A, a lens causes rays of light from a point source (such as a star) to converge to a point called the focus. **B**, rays from an extended source (an object of finite apparent size, such a distant tree) form an inverted image at the focal plane.

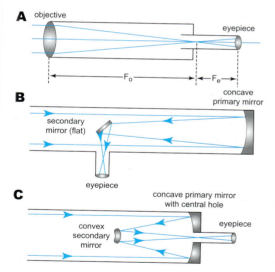

A objective | eyepiece | F_o | F_e | concave primary mirror

B secondary mirror (flat) | eyepiece

C eyepiece | concave primary mirror with central hole | convex secondary mirror | eyepiece

Figure 11.4 The paths followed by light rays in three types of telescope: **A**, a refractor; **B**, a Newtonian reflector; and **C**, a Cassegrain reflector. The image formed at the focus of the objective or the primary mirror, can be magnified with an eyepiece. Fo and Fe denote the focal lengths of the objective and eyepiece, respectively.

through the telescope to its apparent size when viewed directly without optical aid) is equal to the focal length of the objective divided by the focal length of the eyepiece; for example, if a telescope has an objective with a focal length of 1 metre (1000 mm) and an eyepiece with a focal length of 20 mm, the magnifying power will be 1000÷20 = 50 (usually written as 50×).

One problem that afflicts refractors is **chromatic aberration**, the inability of a simple lens to bring all wavelengths of light to a focus at the same point. Because short-wave (blue) light is refracted more than long-wave (red) light, red light is focused further from the lens than blue light and the resulting image of a star is surrounded by a halo of false colour. The problem can be greatly reduced, but not completely eliminated, by using an **achromatic lens** – a lens that consists of two components, each made of different types of glass with different refracting properties. An achromatic lens is capable of bringing two colours to the same focus and greatly reducing the spread of points at which the other colours are focused.

A completely different solution to this problem, first achieved in practice by Isaac Newton, and independently by Jean Cassegrain and James Gregory in the latter part of the seventeenth century, was the invention of the **reflecting telescope**, or **reflector**. This type of telescope employs a concave mirror to collect rays of light and bring them to a focus. Because all rays of light, no matter what their wavelength, are reflected at the same angles from the *front* surface of a mirror, they all arrive at the same focal point and chromatic aberration does not arise.

In a reflecting telescope rays of light from a distant star travel down the telescope tube (which is open at the front end) and are then reflected from the concave surface of the **primary mirror**, converging to form an image at a focal point that lies in front of the mirror. In a **Newtonian reflector**, the converging cone of light is reflected by a small, flat **secondary mirror** to the side of the telescope tube, where an eyepiece or a camera can be placed. In the **Cassegrain telescope**, a convex secondary mirror, placed in front of the focus, reflects the cone of light back down the telescope tube and through a central hole in the primary mirror to a focal point behind the primary mirror. Because the curved secondary mirror changes the angle at which the rays are converging, it also increases the effective focal length of the telescope; this allows a long focal length to be achieved with a relatively short telescope tube (Fig. 11.4). The Cassegrain system, and variants on its basic design, is very widely used in present-day telescopes.

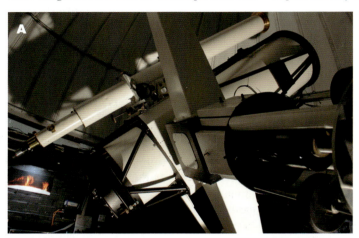

Figure 11.5 Telescopes at Bayfordbury Observatory: **A**, a 0.5-metre aperture Cassegrain reflector with a 16.5-cm refractor mounted on top. **B**, a 0.4-metre Schmidt-Cassegrain telescope with a CCD camera attached. Image credits: Courtesy of Bayfordbury Observatory, University of Hertfordshire.

Hybrid telescopes, such as the Schmidt-Cassegrain and Maksutov, feature a specially shaped thin lens, placed at the front of the telescope tube, which, in association with a concave primary mirror, makes up a compact optical system that is widely used in small and medium-sized telescopes.

Light-gathering and resolution

The two most important functions of a telescope are its **light-gathering power** (its ability to collect light and thereby to reveal objects too faint to be seen with the naked eye) and its **resolving power**, or **resolution** (its ability to distinguish points of light that are too close together to be seen separately by the unaided eye, and to reveal fine detail in the images of distant objects).

Light-gathering power depends on the total surface area of the objective lens or primary mirror and (because the area of a circular lens depends on the square of its diameter) is proportional to the square of the aperture. For example, if one telescope has an aperture twice as great as another, the larger instrument will collect $2 \times 2 = 4$ times as much as the smaller, and will be able to reveal stars four times fainter. The pupil of the human eye, once it has adapted to darkness, expands to an aperture of about 7 mm. Consequently, a telescope with an aperture of 70 mm (ten times greater) should be able to reveal stars 100 times fainter than the unaided eye can see. The largest optical telescopes in the world have apertures in the region of 10 metres (10,000 mm) and so can collect something like two million times as much light as the unaided human eye.

Resolving power can be defined as the minimum angular separation between two identical stars that can *just* be distinguished as separate points of light. Resolving power is inversely proportional to aperture: the greater the aperture, the smaller the angular separation at which close pairs of stars can be distinguished. In practice, large telescopes on the Earth's surface cannot attain their theoretical resolving powers because atmospheric turbulence smears out the incoming light, causing the resulting image to shimmer and shake. Modern telescopes use sophisticated techniques to substantially reduce or virtually cancel out the effects of the atmosphere. One very effective technique is **adaptive optics**, whereby light from the primary mirror is directed onto a small deformable mirror (or mirrors), the shape of which can be modified extremely rapidly to take out the distortions in incoming light waves that have been caused by the atmosphere. These distortions are measured by monitoring a bright reference star (if one is visible in the field of view of the telescope) or an artificial star generated by shining a powerful laser beam into the upper atmosphere (Fig. 11.6). By such means some telescopes can be made to perform almost as well as if the atmosphere were absent.

Currently, the largest optical telescopes in the world are the 10.4-metre aperture Gran Telescopio Canarias on La Palma, and the twin 10-metre Keck telescopes on Mauna Kea, Hawaii. Hopefully, by the early 2020s, these will be exceeded by the Thirty Metre Telescope, due to be erected on Mauna Kea, Hawaii, and by the European Extremely Large Telescope (E-ELT), a 39-metre instrument to be constructed on Cerro Armazones, Chile (Fig. 11.7).

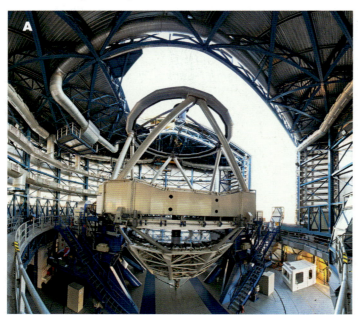

Figure 11.6 The European Southern Observatory's Very Large Telescope (VLT) at Paranal, Chile: **A,** Inside view of one of the four 8.2-metre VLT unit telescopes, with the dome open. Image credit: ESO/B. Tafreshi (twanight.org). **B,** the domes housing the four 8.2-metre unit telescopes(UTs). A laser beam is projecting from UT4 to create an artificial guide star, which is part of the adaptive optics system. Image credit: ESO/S. Brunier.

Figure 11.7 Artist's impression of the 39-metre aperture European Extremely Large Telescope (E-ELT) which is to be constructed on Cerro Armazones, a 3060-metre mountain top in Chile's Atacama desert, and which will become the largest optical/infrared telescope in the world. Image credit: ESO/L. Calçada.

Atmospheric windows

When it is not cloudy, the atmosphere is reasonably transparent to wavelengths in the visible region and a little way beyond, into the near ultraviolet and the near infrared, so we can study objects emitting those wavelengths from ground-based telescopes. This range of wavelengths is called the **optical window**. Most other wavelengths are absorbed at different heights in the atmosphere, or reflected back into space, and cannot be studied from ground level, although several different wavebands in the infrared are accessible at high mountain sites. In the microwave and radio region of the electromagnetic spectrum there is another 'window', called the **radio window**, which allows wavelengths from about a centimetre to around 10 metres to penetrate to ground level, where they can be collected and studied by **radio telescopes**. The most familiar kind of radio telescope is the steerable dish (the most famous being the 80-metre diameter 'Lovell Telescope' at Jodrell Bank in Cheshire, England) (Fig. 11.8). Radio waves falling on a concave metallic surface are focused onto a radio receiver placed at the focus of the dish, and the dish can be swivelled and tilted to point to any location on the sky. Some dishes have the receiver at the prime focus of the dish, whereas others use a system similar to the Cassegrain optical telescope.

Interferometers and telescope arrays

For a telescope of a given aperture, resolution is inversely proportional to wavelength: the longer the wavelength, the poorer the

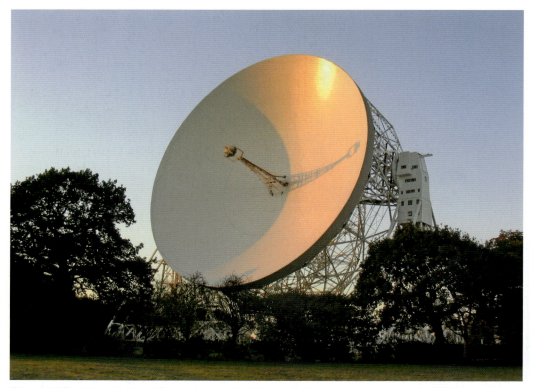

Figure 11.8 The 80-metre Lovell Telescope at Jodrell Bank, UK. Originally commissioned in 1957, it remains the third-largest fully-steerable radio telescope in the world. Image credit: Anthony Holloway, Jodrell Bank.

resolving power. A radio telescope studies wavelengths which, typically, are 100,000 –1,000,000 times longer than visible light. Consequently, to attain the same theoretical resolving power as a 1-metre optical telescope, a single radio dish would have to be 100 to 1000 kilometres in diameter. A solution to this problem is provided by the radio **interferometer** which, by combining the signals received by two radio dishes separated by a particular distance (called the baseline), can attain a resolution (along the direction in which the baseline is orientated) equivalent to that which would be attained by a single dish with an aperture equal to the separation between the two dishes. A more elaborate technique, called **aperture synthesis**, involves conducting a series of observations at different separations while making use of the Earth's rotation to change the orientation of baseline joining the dishes. By this means (though it takes a long time to assemble all the data) an 'image' of a radio source can be constructed that has a resolution equivalent to that which would be obtained by a complete single dish equal in aperture to the maximum separation of the dishes. The process of building up the image can be accomplished more quickly by using a set of moveable dishes mounted on a railway track (or tracks) and varying their separations in a systematic fashion. **Very long baseline interferometry** (**VLBI**) utilizes dishes separated by hundred or thousands of kilometres to achieve resolutions in radio astronomy that can be markedly superior to those attained by the largest conventional optical telescopes.

As with optical telescopes, the greater the collecting area of a dish, or a linked array of dishes, the more sensitive it is and the fainter the radio sources that can be detected. The trend in radio astronomy today is to construct arrays consisting of very large numbers of small dishes rather than single large steerable

ones; they are much cheaper and easier to construct (Fig. 11.9). One example is the Allen Array, which is being used, among many other things, to look for evidence of extraterrestrial signals (*see* Chapter 10). In its present form it consists of 42 dishes of 6-metre aperture, but, when complete, will comprise a total of 350 dishes. The largest current project of this kind is the impending construction of the Square Kilometre Array, which will consist of sets of dishes, located in Africa and Australia, which together will provide a total collecting area of about 1 square kilometre.

Although similar approaches can, in principle, be applied to optical telescopes, so far optical interferometry and aperture synthesis has only been achieved with telescopes in

Figure 11.9 The Atacama Large Millimetre/submillimetre Array (ALMA) is located at an altitude of 5000 metres in one of the driest places on Earth. It consists of 66 individual radio dishes. Image credit: ESO.

close proximity to each other – for example, the four 8.2-metre telescopes, which together with four moveable 1.8-metre telescopes, make up the European Southern Observatory's Very Large Telescope (the VLT).

Seeing beyond the visible

The only way to gain access to the full range of wavelengths arriving from distant sources is to place instruments and telescopes on spacecraft that orbit the Earth, or travel through interplanetary space, and which, therefore, are clear of the obscuring effects of the atmosphere. Over the past few decades, orbiting telescopes (or 'observatories') have accessed all kinds of electromagnetic radiation from gamma rays to infrared and beyond. Such instruments have allowed us to study phenomena as diverse as gamma ray bursts from the most energetic explosions in the universe, exceedingly remote galaxies and protogalaxies whose radiation has been red-shifted far into the infrared region of the spectrum, and the cosmic microwave background radiation. Optical astronomy, too, has benefited hugely from orbiting telescopes such as the Hubble Space Telescope (Fig. 11.10).

Figure 11.10 The 2.4-metre aperture NASA/ESA Hubble Space Telescope in orbit, 600 km above the Earth. Image credit: European Space Agency.

Energetic particles

Information about remote cosmic objects can be gleaned from energetic particles called **cosmic rays** – protons, atomic nuclei and other subatomic particles that have been accelerated to very high fractions of the speed of light by violent processes such as supernovae and hypernovae. When cosmic ray particles plunge into the atmosphere, they produce sprays of secondary particles, some of which can reach ground-based detectors. As energetic cosmic ray particles slow down on entering the atmosphere (they enter the atmosphere at very nearly the vacuum speed of light and so, initially, are travelling faster than the speed of light in air) they emit bursts of light that can be imaged by ground-based instruments. Ground-based detectors can also record flashes that are produced when highly energetic gamma rays plough into the upper atmosphere, and this provides a way of studying cosmic gamma-ray sources from the ground.

Direct exploration of the solar system by spacecraft

Spacecraft orbiting the Earth or travelling through interplanetary space can detect radiation and bursts of subatomic particles emanating from events such as solar flares and can directly sample and measure the solar wind and interplanetary magnetic field. They can also collect, analyse and return to Earth samples of interplanetary dust and microscopic meteoroids. Most significantly, since the early 1960s a diverse armada of spacecraft has flown past each of the eight planets, entered orbit around five of them (Mercury, Venus, Mars, Jupiter and Saturn), landed on the surfaces of the Moon, Venus and Mars, plunged into the atmosphere of Jupiter, flown through the rings of Saturn, landed on the surface of Saturn's massive moon, Titan, touched down on asteroids and impacted the nucleus of a comet (Fig. 11.11). Missions such as these have sent back high-resolution images of planets, their satellites, asteroids

Figure 11.11 This artist's impression of the Rosetta cometary probe shows the main body of the spacecraft, with the smaller Philae lander attached, the communications antenna and the 'wings' of solar panels. Image credit: ESA – J. Huart.

and comets, monitored planetary weather systems, and analysed planetary atmospheres. Landers have imaged geological structures, and investigated the composition of surface rocks and soils on Mars, Venus and Titan, while roving vehicles have explored the surfaces of the Moon and Mars (Fig. 11.12). By these means, our knowledge and understanding of the Solar System have been utterly transformed.

Imaging the Universe

Up until the invention of photography in the mid-nineteenth century, all astronomical observations were carried out by the human eye. Photography brought great advances; permanent images of stars, clusters, nebulae, galaxies and planets could be obtained by placing photographic plates, or film, at the focal plane of telescopes, and longer exposures accumulated photons arriving from faint sources, thereby revealing objects fainter than the human eye could see. The development of photometers (devices that could measure the brightness of objects in a quantitative way) brought new levels of precision to the measurement of brightness and the study of variable sources of light.

The greatest revolution of recent decades has been the development of electronic imaging devices, notably **charge-coupled devices** (**CCDs**), which are arrays of tiny light-sensitive elements (or 'pixels'), each of which generates and stores electrical charges in proportion to the amount of light that falls upon it during an exposure. These charges can then be read off sequentially, to create an image. With their ability to measure brightness accurately, create high-resolution images, and detect the feeble flow of photons arriving from remote stars and galaxies, CCDs have transformed astronomical imaging. The CCD arrays used on some of the world's largest telescopes contain prodigious numbers of pixels, the largest having more than a gigapixel (a billion pixels). The prodigious amounts of data produced by electronic detectors and imaging devices are analysed and processed by computer software, and most professional astronomers today spend little, if any, time actually looking through telescopes. Instead, they book telescope time, acquire their data electronically, and spend much time sitting in front of computer screens interpreting the results.

Amateur astronomers still observe by eye, although many use CCD cameras to produce images of extraordinary sharpness and quality. The ready availability of powerful and compact off-the-shelf telescopes, which can be set up and aligned quickly to point to and precisely follow particular objects as they track across the sky, has revolutionized amateur astronomy. Yet, despite all this sophistication, there is still enormous pleasure to be derived, and real scientific results to be obtained, from visual observation, with or without optical aid – for example, monitoring meteors and meteor showers or observing aurorae, monitoring variable stars, searching for comets, or charting the changing cloud patterns on Jupiter. Amateur observers – some using automated telescopes and imaging systems – also have an impressive track record in discovering novae and supernovae.

Despite computer analysis and image recognition software, the human eye remains the most effective instrument for tasks such as classifying galaxies according to their

Figure 11.12 Self portrait of NASA's Curiosity Rover at the location near the base of Mount Sharp (right) in Gale crater, Mars, where it scooped up its first soil samples. Image credit: NASA/JPL-Caltech/Malin Space Science Systems.

appearance and picking out unusual and anomalous objects. The Galaxy Zoo project (one of a number of citizen science projects that operate under the general umbrella of zooniverse.org), to which anyone with an interest and a computer can sign up, distributes batches of galaxy images that have been obtained by large survey instruments to willing participants, who then undertake to classify them into basic types. This network of citizen scientists has achieved great success (and continues so to do) in classifying vast numbers of individual galaxies more effectively than current computer software can do.

In all of these ways, astronomy remains, as it always has been, a science in which amateurs can and do make real, important and sometimes highly significant contributions.

Glossary

absolute magnitude [68]**:** the apparent magnitude a star would have if it were at a distance of 10 parsecs.

absorption line [69]**:** a dark line in a **spectrum**, which is caused by the absorption of light at a particular wavelength by atoms or **ions** of a particular chemical element.

absorption nebula [82]**:** an interstellar cloud which contains so much dust that it obscures background sources of light and appears as a dark patch against the starry background. Also known as a **dark nebula**.

accretion [85]**:** the accumulation of matter by a massive body under the action of gravity; the colliding and sticking together of small particles or bodies to make progressively larger ones.

accretion disc [93]**:** a disc of gas around a star or a compact massive object that has been accreted from a neighbouring star or gas cloud.

achromatic lens [137]**:** a lens consisting of two components, made of different types of glass, which reduces the effects of **chromatic aberration**.

active galactic nucleus [107]**:** the compact, variable and highly luminous core of an **active galaxy**.

active galaxy [106]**:** a galaxy that radiates up to 10,000 times as much energy as a conventional galaxy such as the **Milky Way**, over a wide range of wavelengths, much of which originates from an extremely compact nucleus (an **active galactic nucleus**).

active region [38]**:** an area of concentrated magnetic fields on the surface, or in the atmosphere, of the Sun which is associated with the occurrence of various forms of **solar activity**.

adaptive optics [139]**:** an optical system that senses and compensates for distortions induced in light waves as they travel through the atmosphere.

altitude [14]**:** the angle between an observer's horizon and a celestial body, measured perpendicular to the horizon.

amplitude [134]**:** the range in size of a varying quantity, for example, the maximum height of a wave above its mean level.

Andromeda galaxy [5]**:** the nearest spiral galaxy, which is located at a distance of 2,500,000 light years in the constellation of Andromeda; otherwise known as M31.

annual parallax [66]**:** the maximum angular displacement of a star from its mean position in the sky due to **parallax**.

annular eclipse [27]**:** an **eclipse** of the Sun during which the dark disc of the Moon is surrounded by a narrow ring of sunlight.

antibaryon [115]**:** the **antiparticle** of a **baryon**.

antiparticle [115]**:** an elementary particle that has the same mass as an ordinary matter particle, but which has opposite (or 'mirror image') values of other properties such as electrical charge or spin.

aperture [136]**:** the clear diameter of the **objective** lens or **primary mirror** of a telescope.

aperture synthesis [142]**:** a technique that enables two or more moveable radio dishes to build up an image equivalent to that which would be attained by a much larger single dish.

aphelion [21]: the point in its orbit around the Sun at which a body is at its greatest distance from the Sun.

apogee [27, 31]: the point in its orbit around the Earth at which a body is at its greatest distance from the Earth.

apparent magnitude [65]: the apparent brightness of a celestial object as seen from the Earth, expressed on the **magnitude scale**.

apparent solar day [17]: the time interval between two successive **upper transits** of the Sun across an observer's **meridian** (i.e., between two successive noons); it is divided into 24 hours of **apparent solar time**.

apparent solar time [17]: time based on the angle between an observer's **meridian** and the Sun.

arcsec [66]: abbreviation for arc second (a second of angular measurement), which is 1/3,600th part of one degree.

asteroid [59]: a small body, otherwise known as a **minor planet**, which revolves round the Sun in an independent orbit.

astrometric binary [75]: a **binary** system in which periodic variations in the motion of a visible star reveal the presence of an unseen companion star.

astronomical unit [22]: a unit of measurement equal to the **semi-major axis** of the Earth's orbit (149,600,000 km); abbreviation, AU.

aurora [44]: varying pattern of light radiated by atoms and ions in the upper atmosphere, stimulated by an influx of charged particles, usually in the vicinity of the north and south magnetic poles.

autumnal equinox [13]: the point of intersection between the **ecliptic** and the **celestial equator** at which the Sun passes from north to south of the celestial equator. The Sun arrives at this point on or around 22 September each year.

azimuth [14]: the angle, measured parallel to the horizon, between a star and a point on the horizon that is vertically below the star.

B

barred spiral galaxy [102]: a **galaxy** in which the **spiral arms** emerge from the ends of a luminous bar or elongated ellipsoid that straddles its **nucleus**.

baryon [115]: a particle, composed of three **quarks**, which is acted on by the strong nuclear force; examples include **protons** and **neutrons** – the building blocks of ordinary matter.

baryonic matter [117]: matter composed of **baryons**.

Big Bang (theory) [113]: the theory which suggests that the **universe** originated about 13.8 billion years ago by expanding from an exceedingly hot, dense initial state.

Big Chill [122]: the dark, cold state towards which an ever-expanding universe may eventually evolve.

Big Crunch [120]: the final state of the universe if it eventually collapses on itself.

Big Rip [122]: the hypothesized catastrophic tearing apart of the universe and all that it contains that might occur if the repulsive influence of dark energy were to increase without limit.

binary (star) [74]: two stars that revolve round each other under the influence of their mutual **gravitation**; specific types of binary include, **astrometric binary**, **spectroscopic binary**, and **eclipsing binary**.

black dwarf [89]: the cold, dark body that is the final state to which, in the distant future, a cooling, fading **white dwarf** eventually will evolve.

black hole [92]: the region of space surrounding a collapsed massive body within which gravity is so powerful that nothing (not even light) can escape.

blazar [110]: the most violently variable type of active galaxy.

blueshift [73, 136]: the observed shortening of the **wavelength** (and increase in **frequency**) of light waves from an approaching source and the resulting displacement of **spectral lines** towards the short-wave (blue) end of the **spectrum**.

brown dwarf [70, 86]: a dense, cool star-like body that is insufficiently massive to enable hydrogen **fusion** reactions to ignite in its core.

C

Cassegrain telescope [138]: a type of reflecting telescope (**reflector**) in which light from the concave **primary mirror** is reflected back from a convex **secondary mirror**, through a central hole in the primary, to a **focus** that lies to the rear of the primary.

celestial equator [10]: the projection of the Earth's equator onto the **celestial sphere**.

celestial meridian [14]: a circle on the celestial sphere that passes through both **celestial poles**, the observer's **zenith** and the north and south points of the horizon (see also, **hour circle**).

celestial pole [10]: one of the two points at which the Earth's axis, extended into space, meets the **celestial sphere**; the north(south) celestial pole is directly above the Earth's north (south) pole.

celestial sphere [10]: an imaginary sphere, of very large radius, which has the Earth at its centre and which rotates round the Earth once a day; in order to specify the positions of celestial objects, it is convenient to think of them as being attached to the inside of this sphere.

central bulge/nuclear bulge [96]: the ellipsoidal distribution of relatively closely spaced stars around the nucleus of a **spiral galaxy**.

centre of mass [74, 123]: the point around which two bodies revolve under the action of their mutual gravitational attraction; it lies on the line joining the centres of the two bodies, closer to the more massive of the two.

Cepheid variable [77]: a type of pulsating variable star that increases and decreases in brightness in a regular periodic fashion; its period is related to its luminosity.

charge-coupled device (CCD) [146]: an electronic imaging device that is divided into a grid of tiny elements each of which builds up an electrical charge proportional to the amount of light that falls upon it.

chromatic aberration [137]: the inability of a simple lens to focus all wavelengths of light at the same point, which arises because different wavelengths are refracted by different amounts (see **refraction**).

chromosphere [35]: the thin layer of the Sun's atmosphere that lies between the **photosphere** and the **corona**.

circular velocity [30]: the speed at which a body (e.g. a satellite) moves if it is travelling in a circular orbit of given radius around a massive body (e.g. a planet).

circumpolar star [11]: a star that remains at all times above the horizon of an observer at a particular latitude on the Earth.

coma [60]: a cloud of gas and dust that surrounds the nucleus of a comet and forms its visible 'head'.

comet [60]: a small body that consists of ice, dust and rocky material, from which streams of gas and dust evaporate to create a **coma** and **tail(s)** each time it makes a close approach to the Sun.

conjunction [22]: a close alignment of two (or more) celestial bodies in the sky.

constellation [8]: a grouping, or pattern, of stars that occupies a particular region on the **celestial sphere**.

continuous spectrum [69]: an unbroken distribution of **electromagnetic radiation** extending over a broad range of wavelength, which at visible wavelengths appears as a rainbow band of colours.

convective zone [34]**:** the region within the Sun, or a star, through which energy is transported predominantly by convection.

coordinated universal time (UTC) [18]**:** a precise system of time measurement, related to **Universal Time**, but which is regulated by a set of atomic clocks.

core [47]**:** the central region of a star or planet, or a localised dense concentration of material within a cloud of gas and dust.

core collapse [89]**:** the abrupt collapse of the core of a massive star, which occurs when it runs out of 'fuel' and cannot sustain further **fusion** reactions, and which leads to the formation of a **neutron star** or a **black hole**.

corona [36]**:** the tenuous, high-temperature, outermost region of the Sun's atmosphere.

coronal hole [42]**:** apparently dark, low-density region in the solar corona, from which solar **plasma** escapes into interplanetary space.

coronal mass ejection [41]**:** a massive bubble of **plasma** which is ejected from the Sun's **corona** and propagates out through interplanetary space.

cosmic microwave background radiation (CMBR) [116]**:** a faint background of microwave radiation that is uniformly distributed across the whole sky (apart from small-scale variations of about one part in 100,000) and which is remnant radiation from the **Big Bang**.

cosmic rays [145]**:** highly energetic subatomic particles (e.g., electrons, protons and heavier atomic nuclei) that travel through space at exceedingly large fractions of the speed of light.

cosmological constant [121]**:** a quantity initially invoked by Einstein to enable the universe to be static; it relates to the concept that space is uniformly permeated by a dilute form of energy called **vacuum energy** (see also **dark energy**).

cosmological redshift [114]**:** the **redshift** in the spectra of galaxies that is caused by the expansion of the universe.

critical density [120]**:** the average density of matter and energy that would make the overall geometry of the universe flat (see **flat universe**); the mean density of a universe which just, but only just, would be able to expand forever if its behaviour were governed by gravity alone.

crust [46]**:** the thin outermost solid layer of a planet or a major planetary satellite, the interior of which has separated into distinct layers (see also, **core**, **mantle**).

culmination/upper transit [15]**:** the position of a star when it is crossing an observer's celestial meridian and is at its greatest altitude.

D

dark energy [117]**:** an unknown form of energy, which makes up about 68% of the total mass and energy content of the universe, and which is believed to be driving the accelerating expansion of the universe.

dark matter [97]**:** matter that emits no detectable radiation, but which exerts a gravitational influence on its surroundings; its nature is unknown, but the front-running hypothesis is that it consists of particles called **WIMPs**.

dark nebula [82]**:** a cloud of gas and dust that contains enough dust to obscure the light of more distant stars; it appears as a dark patch against the starry background.

declination [14]**:** the angle measured north (+) or south (–) between the **celestial equator** and a star or other celestial body.

density fluctuations [116]**:** marginally denser and less dense regions within the primordial mix of matter and radiation that existed at the time when the **cosmic microwave background radiation** was released (see **temperature fluctuations**).

differential rotation [38]**:** the phenomenon

whereby the rotation period at the surface of a body, such as the Sun, has different values at different latitudes.

diffraction grating [135]: a plate on which a large number of parallel grooves has been cut, which acts like a prism to disperse different wavelengths of light into a spectrum.

direct motion [20]: the apparent motion of a celestial body, such as a planet, from west to east relative to the background stars.

dispersion [135]: the spreading out of light into its constituent wavelengths by a lens, prism or **diffraction grating**.

Doppler effect [73, 136]: the observed change in wavelength and frequency of light (or other form of electromagnetic radiation) that is caused by the motion of a source directly towards or away from an observer (the term originally applied to an analogous effect with sound waves).

dwarf planet [58]: a body that orbits the Sun, and which is sufficiently massive for its own gravity to have pulled it into a near-spherical shape, but is not massive enough to have swept all other bodies out of the region in which it is travelling.

E

eclipse [27]: the passage of one celestial body into the shadow cast by another (e.g. an eclipse of the Sun occurs when the Moon passes between the Sun and the Earth, and an eclipse of the Moon occurs when the Moon passes into the Earth's shadow).

eclipsing binary [75]: a **binary** system in which each member star alternately passes in front of the other, cutting off all or part of the light of the other, so causing periodic variations in the combined light of the two stars.

ecliptic [13]: the apparent annual path of the Sun on the **celestial sphere** relative to the background stars; it lies in the plane of the Earth's orbit.

electromagnetic force [117]: the force of attraction or repulsion that acts between electrically charged particles, such as protons and neutrons, and which controls the emission and absorption of electromagnetic radiation.

electromagnetic radiation [134]: a wave-like electric and magnetic disturbance that travels through space at the speed of light.

electromagnetic spectrum [40, 134]: the full range of electromagnetic radiations, from the shortest-wavelength (highest energy) gamma rays to the longest-wavelength (lowest energy) radio waves (see Fig. 11.1).

electron [69]: a lightweight subatomic particle that has negative electrical charge.

ellipse [20]: an oval geometrical figure (the maximum diameter of which is called the **major axis**) that is characterized by two points, called **foci**, which lie on the major axis, on opposite sides of its centre; the greater the separation between the foci, the more elongated the ellipse.

elliptical galaxy [100]: a galaxy that has an apparently spherical or elliptical shape and which generally contains much less gas and dust than a typical **spiral galaxy**.

elongation [23]: the observed angle between the Sun and a planet.

emission line [69, 80]: a bright line in a **spectrum** that corresponds to light of one particular wavelength, and which is emitted when electrons drop down from a higher to a lower energy level within atoms or **ions** of a particular chemical element.

emission nebula [80]: a luminous gas cloud that is ionized by, and stimulated to emit light by, intense ultraviolet radiation emanating from one or more hot, highly-luminous stars that are embedded within it.

escape velocity [31]: the minimum speed at which a projectile must be fired in order to continue always to recede from a massive body and not fall back.

event horizon [93]**:** the boundary of a black hole, which derives its name from the fact that no particle, signal or information of any kind can propagate outwards through this boundary.

exoplanet [4, 123]**:** a planet that revolves around another star (i.e. around a star other than the Sun). Also known as 'extrasolar planet'.

extraterrestrial intelligence (ETI) [131]**:** a term used to describe hypothetical intelligent species that may exist elsewhere in the universe, and which may (or may not) possess the means to communicate over interstellar distances.

eyepiece [136]**:** a lens with a short **focal length** that is used to magnify the image produced by the **objective** or **primary mirror** of a telescope.

F

filament (solar) [38]**:** a cloud of denser gas, suspended in the solar atmosphere, which, when viewed through an appropriate narrow-band filter, shows up as a dark feature against the bright background of the solar disc (see also, **prominence**).

fireball [63]**:** an exceptionally bright **meteor**, usually associated with the passage of a **meteorite** through the atmosphere.

first quarter [26]**:** the half-illuminated **phase** of the Moon that occurs about a week after **New Moon**, when the angle between the Sun and the Moon in the sky is 90°.

flare star [77]**:** a cool, low-luminosity star that undergoes frequent short-lived increases in brightness, typically of a few minutes' duration.

flat universe [120]**:** a universe in which the overall net curvature of space is zero, and the mean density of matter and energy is equal to the **critical density**.

flux [33]**:** the amount of radiant energy per second passing perpendicularly through an area of one square metre.

flux tube [44]**:** a bundle of magnetic field lines.

focal length [136]**:** the distance between the centre of a lens, or the centre of the front surface of a concave mirror, and its focal point.

focal point/focus [136]**:** the point at which rays of light from a distant point source of light, refracted by a lens or reflected by a mirror, intersect to form a point-like image.

frequency [134]**:** the number of wave crests of a wave motion (e.g. light) that pass a given point in one second.

Full Moon [26]**:** the fully illuminated phase of the Moon, which occurs when the Moon is on the opposite side of the Earth from the Sun, and the angle between the Sun and the Moon is 180°.

fusion [33]**:** the process whereby lighter atomic nuclei (see **nucleus**) are welded together to form heavier nuclei with an associated release of energy.

G

galaxy [4, 95]**:** a large aggregation of stars, gas, dust and dark matter that typically contains between a few million and several trillion stars and has a diameter in the range from a few thousand to several hundred thousand light-years.

galaxy cluster [103]**:** a collection of galaxies held together by gravity and containing between a few tens and a few thousands of members.

Galilean satellites [55]**:** the four major moons of the planet Jupiter (Io, Europa, Ganymede and Callisto), which were first observed in 1609–1610 by Italian astronomer, Galileo Galilei.

geocentric system [20]**:** the theory, developed two millennia ago by the ancient Greeks, that the Sun, Moon, planets and stars revolve round a central Earth.

geostationary orbit [30]**:** a circular orbit around the Earth, with a radius of about 42,000 km and in the plane of the Earth's equator, in which a satellite has orbital period equal to the Earth's rotation period and so, when viewed from the Earth's surface, appears to be stationary in the sky.

giant planet/gas giant [46]: a planet that is much larger than the Earth, and which is composed mainly of hydrogen, helium and hydrogen compounds such as methane and ammonia; the four giants of the Solar System are Jupiter, Saturn, Uranus and Neptune.

giant (star) [72]: a star which is substantially larger and more luminous than a **main sequence star** with the same surface temperature.

globular cluster [96]: a near-spherical star cluster that contains between 10,000 and 1,000,000 predominantly old stars.

Goldilocks planet [129]: colloquial name for a planet that revolves within the **habitable zone** of its host star and on which conditions may be suitable for the existence of life.

gravitation (gravity) [29]: the attractive force that acts between material bodies and particles; the force of attraction between two bodies depends on the product of their masses and is inversely proportional to the square of the distance between their centres.

gravitational lensing [106]: the formation of an image, or images, of background objects caused by the deflection of light as it passes close to a massive foreground object or through a distribution of mass such as a galaxy cluster.

gravitational slingshot [31]: a technique whereby a spacecraft can gain (or lose) speed, in its motion relative to the Sun, by passing through the gravitational field of a planet.

gravitational wave [117]: a wave-like disturbance of space that propagates at the speed of light.

greenhouse effect [47]: the effect whereby an atmosphere can raise the surface temperature of a planet by absorbing and re-emitting outgoing infrared radiation that otherwise would escape directly to space.

H

habitable zone [127]: the region around a star within which (depending on the properties of its atmosphere) an **exoplanet**'s temperature would be such that liquid water could exist on its surface. Colloquially called 'Goldilocks Zone'.

halo [96]: an extensive spheroidal distribution of **globular clusters** and old stars (and dark matter) around a **galaxy**.

heliocentric theory [20]: a theory, such as the one that was proposed in 1543 by Nicolaus Copernicus, in which the Earth and planets revolve round a central Sun.

heliopause [32, 43]: the boundary of the **heliosphere**.

heliosphere [43]: the region of space around the Sun within which the **solar wind** and interplanetary magnetic field is confined by the pressure exerted by the **interstellar medium**.

Hertzsprung-Russell diagram [70]: a diagram on which the **absolute magnitude** or **luminosity** of stars are plotted against **spectral class** or **surface temperature**.

hot Jupiter [124]: an **exoplanet** that is comparable in mass to Jupiter (i.e. a **gas giant**), which orbits close to its host star and therefore has a very high temperature.

hour circle [14]: a circle that passes through both **celestial poles** and which is perpendicular to the **celestial equator** (see also **celestial meridian**).

H-R diagram [70]: a commonly used abbreviation for Hertzsprung-Russell diagram.

Hubble classification scheme [100]: a scheme for classifying galaxies according to their appearance that was devised by Edwin Hubble; the principal classes are elliptical, spiral, barred spiral and irregular.

Hubble constant [113]: the constant of proportionality (symbol H) between the speeds at which galaxies are receding and their distances.

Hubble law [112]: the observed relationship between the **redshifts** in the spectra of galaxies and their distances, which implies that the speeds at which galaxies are receding is directly proportional to their distances.

Hubble time [113]: the time that galaxies would have taken to recede to their present distances, if the universe were expanding at a constant rate; it is equal to the reciprocal of the Hubble constant.

hydrogen burning [86]: a colloquial term to describe the conversion of hydrogen to helium by means of **fusion** reactions (it is not 'burning' in the conventional sense).

hypernova [93]: an exceptionally powerful type of **supernova** explosion, possibly triggered when the core of a very massive star collapses to form a **black hole**.

I

impact basin [48]: a very large, crater-like, depression excavated by the impact of a giant **meteorite** or **asteroid**-sized body; it may subsequently have been flooded by magma (molten material).

inferior conjunction [22]: the configuration that occurs when an **inferior planet** passes between the Sun and the Earth; its **elongation** is then 0°.

inferior planet [22]: a planet that is closer to the Sun than is the Earth (i.e. the planets Mercury and Venus).

inflation [114]: the short-lived, but dramatic, exponential expansion of space that is believed to have occurred a tiny fraction of a second after the Big Bang.

infrared (radiation) [134]: **electromagnetic radiation** with **wavelength**s longer than visible light, which lie within the wavelength range 0.7–350 **micrometres** (μm).

interferometer [142]: a device that achieves improved **resolution** by combining the signals received by two or more radio dishes or optical mirrors.

interstellar cloud [80]: a cloud of gas and dust in **interstellar space**.

interstellar dust [82]: microscopic solid grains of matter that exist in **interstellar clouds** and **dark nebulae**.

interstellar matter [80]: gas and dust spread through interstellar space.

interstellar space [43]: the space between the stars.

ion [48]: an atom that has lost one or more of its normal complement of electrons (positive ion) or which has gained one or more extra electrons (negative ion).

ionosphere [44, 48]: a layer, or layers, of **ions** and **electrons** formed in the upper atmosphere by the action of solar **ultraviolet radiation**.

irregular galaxy [103]: a **galaxy** that has no well-defined structure or symmetry; denoted by Irr in the **Hubble classification scheme**.

K

Kepler's laws of planetary motion [20]: three laws, formulated by Johannes Kepler in the seventeenth century, that describe the motion of planets in their elliptical **orbits** around the Sun.

Kuiper belt [60]: a flattened distribution of icy asteroids and cometary nuclei that lies beyond the orbit of Neptune and extends out to about 100 **astronomical unit**s from the Sun.

L

Large Magellanic Cloud [103]: the larger of two satellite galaxies of the Milky Way, which lies at a distance of 163,000 light years, and is visible to the naked eye in the southern constellation Dorado.

last quarter [27]: the half-illuminated **phase** of the Moon that occurs about a week before **New Moon**, when the angle between the Sun and the Moon in the sky is 90°.

latitude [13]: the angular distance north or south of the equator of a place on the Earth's surface.

lenticular galaxy [102]**:** a lens-shaped **galaxy** that has a **central bulge** and disc (see **galactic disc**) but no **spiral arms**; denoted by S0 in the **Hubble classification scheme**.

light curve [75]**:** a graph of the variation with time of the brightness of a light source (e.g. a **variable star**).

light-gathering power [139]**:** a measure of the ability of a telescope to collect light; it is proportional to the square of the telescope's **aperture**.

light-travel time [2]**:** the time taken for a ray of light to travel a given distance at the **speed of light** (300,000 km/s).

light-year [3]**:** a unit of distance equal to the distance travelled by a ray of light in one year (9.46 trillion km).

limb [35]**:** the edge of the visible disc of a celestial body (e.g. the Sun or the Moon).

limb darkening [35]**:** the fading of the observed brightness of the **photosphere** towards the edge of the Sun's visible disc.

liquid metallic hydrogen [54]**:** hydrogen under such high pressure that it becomes ionized and behaves like a liquid metal.

lithosphere [47]**:** the outer layer of the Earth that comprises the crust and the uppermost region of the mantle. It has fractured into a number of **lithospheric plates**.

lithospheric plate [47]**:** a relatively rigid piece of the **lithosphere** that moves relative to neighbouring plates (see also, **plate tectonics**).

Local Group [103]**:** the group of about thirty galaxies to which the **Milky Way** galaxy belongs, the other major members being the **Andromeda galaxy** (M31) and the Triangulum galaxy (M33).

longitude [13]**:** the angle between the **meridian** passing through a particular place and the Greenwich meridian (the meridian that passes through the old Royal Observatory at Greenwich).

luminosity [33, 66]**:** the total amount of energy emitted per second by the Sun, a star, or other source of radiation.

M

magnetic field [36]**:** the region around a magnetized body within which the strength and orientation of its magnetic influence can be detected and measured.

magnetic reconnection [41]**:** the joining together of oppositely directed magnetic field lines, which results in the sudden release of energy (see also, **solar flare**).

magnetic storm [44]**:** a disturbance in the Earth's magnetic field caused by solar outbursts (e.g. **flares** and **coronal mass ejections**) and major fluctuations in the **solar wind**.

magnetopause [48]**:** the outer boundary of a **magnetosphere**.

magnetosphere [43, 48]**:** the region of space around a body within which its magnetic field is confined.

magnetotail [48]**:** the part of a planet's magnetosphere on the opposite side of the planet from the Sun, which has been dragged out into an elongated 'tail' by the solar wind and interplanetary magnetic field.

magnification/magnifying power [136]**:** the ratio of the apparent angular size of an object when viewed through a telescope to its apparent angular size when viewed without a telescope.

magnitude scale [65]**:** a logarithmic scale for quantifying the brightness of a celestial object; each step of one magnitude corresponds to a brightness difference equal to the fifth root of 100 (i.e., a factor of 2.512).

main sequence [72, 86]**:** a band of stars on the **Hertzsprung-Russell diagram** which slopes down from the upper left (high luminosity, high temperature) to the lower right (low luminosity, low temperature).

main-sequence star [72]**:** a star that lies on the main sequence and which is powered by fusion reactions that convert hydrogen to helium in its core.

major axis [20]**:** the longest diameter of an ellipse.

mantle [47]**:** the region, composed of dense rock, which lies between the core and the crust of a **terrestrial planet**.

mare (plural: 'maria') [48]**:** a relatively smooth lava-filled basin on the surface of the Moon; the name derives from the Latin word for 'sea'.

mean solar time/mean time [17]**:** a system of time measurement based on the mean duration of the solar day.

Megaparsec (Mpc) [113]**:** a unit of distance measurement equal to one million **parsecs** and equivalent to 3,260,000 **light-years**.

meridian [131]**:** a circle on the surface of the Earth that passes though the north and south poles and crosses the equator at right angles (see also, **celestial meridian**).

mesosphere [48]**:** the layer in the Earth's atmosphere, between the **stratosphere** and the **thermosphere**, within which the temperature decreases with increasing altitude.

meteor [61]**:** the short-lived trail of light that is seen when an incoming meteoroid is vaporized in the Earth's atmosphere.

meteor shower [61]**:** a display of meteors that appear to radiate from a common point in the sky (the **radiant**), which occurs when the Earth passes through a stream of **meteoroids**.

meteorite [63]**:** a rocky or metallic body that survives its passage through the atmosphere to reach the ground in one piece or in fragments.

meteoroid [61]**:** a particle, or small body, which orbits the Sun in interplanetary space.

Milky Way [4, 95]**:** a faint, misty band of light that crosses the sky and which consists of the combined light of a vast number of stars and luminous nebulae that lie in the plane, and in the spiral arms of the **galaxy** to which the Solar System belongs. The term 'Milky Way' (or 'Milky Way galaxy') is also used as a popular name for our galaxy.

minor planet [59]**:** an alternative term (less commonly used) for **asteroid**.

molecular cloud [84]**:** a cool cloud of hydrogen and helium that is rich in molecules and within which conditions are conducive to the formation of stars.

N

nanometre (nm) [134]**:** a unit of measurement equal to one billionth (10^{-9}) of a metre.

near-Earth object (NEO) [60]**:** an asteroid or comet that follows an orbit which comes close to, or may intersect, the orbit of the Earth. By convention, an object is considered to be an NEO if its **perihelion** distance is less than 1.3 **astronomical units** (see also, **potentially hazardous asteroid**).

nebula [80]**:** a cloud of interstellar gas and dust, which may emit light (see **emission nebula**), reflect light (see **reflection nebula**) or absorb light (see **absorption nebula**); 'nebula' is the Latin word for cloud.

neutron [34, 69]**:** a subatomic particle that has zero electrical charge and a mass fractionally greater than that of the **proton** (see also, **baryon**).

neutron star [68, 90]**:** the collapsed core of a high-mass star, which has been compressed to extreme density, and which is composed almost entirely of closely packed neutrons (see also, **pulsar**).

New Moon [25]**:** the stage in its cycle of **phases** at which the Moon is at its closest to the Sun in the sky and its Earth-facing hemisphere is not illuminated.

Newtonian reflector [138]**:** a type of reflecting telescope in which the converging cone of light from the primary mirror is reflected to an eyepiece at the side of the telescope tube.

node [24]**:** one of two points at which an orbit crosses some other reference plane (usually the plane of the Earth's orbit (the ecliptic).

nova [78]**:** a star that experiences a sudden increase in brightness by a factor of between a thousand and a million, but which eventually fades back to its pre-outburst brightness.

nucleus (of an atom) [33, 69]**:** the central core of an atom, which is composed of **protons** and **neutrons** (except in the case of a normal hydrogen atom, where the nucleus consists of a single proton).

nucleus (of a comet) [60]**:** the compact solid core of a comet, which consists of ice, dust and rocky material.

nucleus (of a galaxy) [96]**:** the compact core of a galaxy, where stars are densely concentrated, and which may contain a **supermassive black hole**.

O

objective/object glass [136]**:** The lens that collects light and forms images in a refracting telescope (see **refractor**).

observable universe [7]**:** that part of the universe which, in principle, can be detected from the Earth.

Oort cloud [61]**:** a cloud of icy **planetesimals** and **cometary nuclei** that surrounds the Solar System and extends to a radius of about a light-year, and from which long-period and 'new' comets originate.

opposition [24]**:** the configuration that occurs when the elongation of a superior planet is 180°; it is then on the diametrically opposite side of the sky from the Sun and culminates at midnight (see **culmination**).

optical window [141]**:** a range of wavelengths from the near ultraviolet, through visible light, to the very near infrared, to which the Earth's atmosphere is reasonably transparent.

orbit [20]**:** the path of a body that is revolving round another under the influence of gravity.

orbital period [21]**:** the period of time during which a body travels once around its orbit (see also, **sidereal period**).

P

parallax [65]**:** the apparent shift in the position of an object that occurs when it is observed from two different locations.

parsec [66]**:** a unit of distance that is defined as the distance at which a star would have an **annual parallax** of 1 arcsec; it is equal to 206265 **astronomical units**, or 3.26 **light-years**.

partial eclipse [27]**:** an **eclipse** in which one body is partly obscured by another, or in which only part of one celestial body passes into the cone of dark shadow cast by another.

penumbra [27, 36]**:** (1) the outer part of the shadow cast by a body such as a planet, from within which an observer would see a **partial eclipse** of the Sun; (2) the less cool and less dark outer part of a **sunspot**.

perigee [27, 31]**:** the point on its orbit or trajectory at which a body (e.g., the Moon) is at its closest to the Earth.

perihelion [21]**:** the point in its orbit or trajectory at which a planet, or other body, is at its closest to the Sun.

period–luminosity relationship [98]**:** a relationship between the luminosities of pulsating variable stars (notably **Cepheid variables**) and their periods of variation: the higher the luminosity, the longer the period.

phase [25]**:** the fraction of the Earth-facing hemisphere of the Moon or a planet that is illuminated by the Sun at a particular time.

photon [34, 70, 134]**:** a quantum (a discrete package) of electromagnetic radiation, which can be regarded as a 'particle' of light. The energy of a photon is inversely proportional to its associated wavelength.

photosphere [33]**:** the visible 'surface' of the Sun, which is the layer at the bottom of the Sun's atmosphere from which visible light escapes into space.

planet [1, 20]**:** a body that revolves around the Sun or another star, and shines by reflecting the light of its host star. As a general rule of thumb, a body is considered to be a planet if its mass is greater than that of a **dwarf planet** but less than about 13 times the mass of Jupiter.

planetary nebula [88]**:** a luminous shell of gas that has been ejected from a star at a late stage in its evolution.

planetary system [85]**:** a set of **planets** that revolve round a particular star (see also, **exoplanet**).

planetesimals [85]**:** small solid bodies, 5–10 km in diameter, composed of dust, rock or ice, from which the planets were assembled by the process of **accretion**.

plasma [36]**:** an ionized gas that consists of equal numbers of positively-charged ions, or atomic nuclei, and negatively-charged electrons.

plate tectonics [47]**:** the theory which implies that major geological structures on the Earth are created by the relative movement of **lithospheric plates** (see also, **lithosphere**).

positron [115]**:** an elementary particle that has the same mass as an **electron** but has positive (rather than negative) electrical charge; the **antiparticle** of the electron.

potentially-hazardous asteroid (PHA) [60]**:** an asteroid that can approach closer to the Earth than 0.05 astronomical units, and which has a finite chance of colliding with our planet.

precession [15]**:** the slow change in the orientation of the Earth's axis of rotation which, over a period of 25,800 years, causes the positions of the **celestial poles** to trace out circles on the **celestial sphere**.

primary mirror [138]**:** the main mirror in a reflecting telescope, or **reflector**, which collects light and forms an image.

prominence [38]**:** a plume, or suspended cloud, of gas in the solar atmosphere seen beyond the visible **limb** of the Sun. Active or eruptive prominences undergo rapid changes, whereas quiescent ones may show little overall change for weeks on end.

proper motion [73]**:** the slow annual angular shift in a star's position, caused by its **transverse velocity**.

protogalaxy [118]**:** a cloud of material that was a precursor to the formation of a **galaxy**.

proton [33, 69]**:** a heavy subatomic particle that has positive electrical charge and which is a constituent of every atomic nucleus; it is a **baryon** that is composed of three **quarks**.

proton–proton reaction [34]**:** the series of nuclear reactions by means of which hydrogen nuclei are fused together to form nuclei of helium; the dominant 'hydrogen-burning' reaction in stars comparable to, or less massive than, the Sun.

protostar [86]**:** a star in the early stages of formation, which is still contracting and within which nuclear fusion reactions have not yet commenced.

pulsar [91]**:** a source that emits short pulses of radio waves (or other forms of **electromagnetic radiation**) at very precise intervals; believed to be a rapidly-rotating **neutron star**.

pulsating variable [77]**:** a star that varies in luminosity and temperature while expanding and contracting in a periodic fashion.

Q

quark [115]**:** a fundamental particle that binds together in groups of three to form **protons**, **neutrons** and other types of **baryon**.

quasar [110]**:** an extremely compact and highly luminous **active galactic nucleus**. Although the term is an abbreviation for quasi-stellar *radio* source, it is also commonly applied to **quasi-stellar objects** (QSOs), which have similar characteristics but are not strong radio sources.

quasi-stellar object (QSO) [110]**:** an extremely compact and highly luminous **active galactic nucleus**, which radiates **electromagnetic radiation** across a wide range of wavelengths. Only about ten percent of these objects are 'radio loud' (i.e., are very powerful emitters of radio waves); see **quasar**.

R

radial velocity [73, 124]**:** the component of a body's velocity that is in the radial direction, directly towards or away from the observer (*see* also, **transverse velocity**).

radiant [61]**:** the point on the **celestial sphere** from which the tracks of meteors that are members of a **meteor shower** appear to radiate.

radiation pressure [60]**:** the small, but finite, pressure exerted when **photons** (of light, or other types of **electromagnetic radiation**) collide with particles of matter, or a surface.

radiative zone [34]**:** the region inside a star through which energy is transported by **electromagnetic radiation** (the outward flow of **photons**).

radio galaxy [110]**:** a type of active galaxy that is a powerful emitter of radio waves.

radio telescope [141]**:** an instrument that is designed to detect radio sources; the most familiar type is a concave dish that reflects radio waves to a detector located at its **focus**.

radio window [141]**:** a range of wavelengths from a few centimetres to about 10 metres, to which the Earth's atmosphere is transparent.

radius vector [21]**:** the line directed from the centre of the Sun (or from any body round which another is orbiting) to an orbiting body such as a planet (*see* also **Kepler's laws of planetary motion**).

red giant [72]**:** a star of high luminosity and low surface temperature, which has evolved away from the main sequence and which is very much larger than a main-sequence star with the same surface temperature.

redshift [73, 112, 136]**:** the displacement of lines in the spectrum of a receding source of light to wavelengths longer than would be the case if the source were stationary (*see* also, **Doppler effect**).

red supergiant [68]**:** an extremely luminous and exceptionally large low-temperature star.

reflecting telescope/reflector [137]**:** a telescope that uses a concave mirror to collect light and bring it to a **focus**.

reflection nebula [83]**:** a cloud of dust that reflects light from a nearby star or stars.

refracting telescope/refractor [136]**:** a telescope that uses a lens to collect light and form an image.

refraction [135]**:** the deflection of a ray of light that occurs when it crosses the boundary between two different substances, for example, when passing from air into glass.

resolving power/resolution [139]**:** a measure of the ability of a telescope to reveal fine detail in the image of a distant object; the minimum angular separation between two identical stars that can *just* be distinguished as separate points of light.

retrograde motion [20]**:** the apparent motion of a planet in an east to west direction relative to the background stars (*see* also, **direct motion**).

right ascension [14]**:** the angle between the **hour circle** that passes through the **vernal equinox** and the hour circle that passes through a particular celestial object, measured eastwards from the vernal equinox (*see* also, **declination**).

S

satellite [1, 48]**:** a smaller body (natural or artificial) that orbits around a planet; a 'moon'.

Schwarzschild radius [93]**:** the radius within which a body of particular mass must be compressed in order that light will not be able to escape from it; the radius of a (non-rotating) black hole.

secondary mirror [138]**:** a small flat or curved mirror that deflects the converging cone of light from the **primary mirror** of a **reflecting telescope** to a convenient point at which an eyepiece or detector may be placed; a curved secondary will alter the effective focal length of the instrument.

semi-major axis [21]: half of the **major axis** of an ellipse (i.e. the distance from the centre of the ellipse to either end of its major axis; in the context of **Kepler's laws of planetary motion**, a planet's mean distance from the Sun.

Seyfert galaxy [110]: a type of spiral, or barred spiral, galaxy that contains a particularly bright, compact **nucleus**; a class of **active galaxy**.

sidereal day [17]: the time interval between two successive upper transits of the vernal equinox, which is equal to the rotation period of the Earth relative to the background stars. It is divided into 24 hours of **sidereal time**.

sidereal period (orbital) [25]: the time that one body takes to make one complete circuit of its orbit around another body (e.g., a planet round the Sun), measured relative to the background stars.

sidereal time [17]: a time system based on the rotation of the Earth relative to the background stars.

singularity [93]: a point at which matter is compressed to infinite density; the central point in a (non-rotating) **black hole**.

Small Magellanic Cloud [103]: the smaller of two irregular satellite galaxies of the Milky Way; located at a distance of 200,000 light-years, it is visible to the naked eye in the southern constellation of Tucana.

solar activity [43]: variable, magnetically driven, phenomena that occur on the surface, and in the atmosphere, of the Sun.

solar constant [33]: the average amount of solar energy passing perpendicularly through an area of one square metre in one second at the top of the Earth's atmosphere; its value is about 1.37 kW/m^2.

solar cycle [43]: the cyclic variation in **solar activity** (e.g. the number of **sunspots**, **flares** and **coronal mass ejections**) that reaches a maximum at intervals of about 11 years.

solar flare [40]: a sudden, violent release of up to 10^{25} J of energy (in the form of energetic particles and electromagnetic radiation), which takes place in the solar atmosphere above a complex **active region**.

Solar System [1]: the system of bodies that consists of the Sun, together with the planets, dwarf planets and minor bodies that revolve around it.

solar wind [42]: the outward flow of charged particles (mainly electrons and protons) that emanates from the Sun's **corona**.

space weather [43]: variations in the Earth's space environment, caused by fluctuations in the **solar** wind and interplanetary **magnetic field**, and by bursts of energetic particles and radiation emanating from **solar flares** and **coronal mass ejections**.

spectral class/type [70]: a classification of stars based on the characteristics and relative strengths of the various different lines in their **spectra**. The principal classes are labelled, in descending temperature order, as O, B, A, F, G, K and M.

spectroscope/spectrograph [69, 135]: a device that spreads out the different wavelengths of light into a **spectrum**, which may then be viewed directly with an eyepiece (spectroscope) or recorded photographically or electronically (spectrograph).

spectroscopic binary [75]: a **binary** in which, although the two component stars are too close together to be resolved, its binary nature can be deduced from the fact that the **spectrum** of what appears to be a single star contains two sets of lines that exhibit periodic changes in wavelength as the two stars revolve round each other.

spectrum [69, 135]: a plot of intensity (brightness) against wavelength, obtained when a beam of light (or other form of **electromagnetic radiation**) is spread out into its constituent wavelengths; the spectrum of white light (sunlight) appears to the human eye as a rainbow band of colours.

speed of light [34, 134]**:** the speed at which light propagates. The speed of light in a vacuum, denoted by the symbol, **c**, is constant, and is equal to 299,792,485 metres per second; in round figures, it is 300,000 km/s. Light travels more slowly in a medium such as air, water or glass.

spiral arms [97]**:** lanes in which gas, dust, star-forming regions and hot, highly luminous young stars are concentrated, and which appear to extend out in a spiral pattern from the **central bulge** of a spiral or barred spiral **galaxy**.

spiral galaxy [100]**:** a **galaxy** in which a pattern of **spiral arms** composed of gas, dust and hot young stars spreads out from its **central bulge**.

sporadic meteor [61]**:** a **meteor** that appears at random and which is not associated with a **meteor shower**.

standard candle [99]**:** a high-luminosity object embedded within a distant galaxy (for example, a **Cepheid variable** or a **supernova**), the luminosity of which can be estimated with some confidence; its distance can be determined by comparing its apparent brightness with its assumed inherent luminosity.

standard ruler [99]**:** an object of finite angular size (for example an **emission nebula** embedded within a distant galaxy), the diameter of which can be estimated with some confidence; its distance can be determined by comparing its apparent size with its assumed physical size.

star [1]**:** a self-luminous gaseous body, such as the Sun.

stellar wind [84]**:** an outflow of atomic nuclei and electrons from the atmosphere of a star (see also **solar wind**).

stratosphere [48]**:** the layer within the Earth's atmosphere, which extends from an altitude of about 12 km to around 50 km, within which temperature rises with increasing altitude.

strong nuclear force [117]**:** the force that holds protons and neutrons together in the nucleus of an atom, and that binds together the quarks, which are the internal constituents of **baryons**.

summer solstice [15]**:** the point on the ecliptic at which the Sun is at its greatest northerly **declination**; the Sun reaches this point on or around 21 June each year.

sunspot [36]**:** a patch on the **photosphere** that appears dark because it is cooler than its surroundings. Sunspots are associated with concentrated localized magnetic fields (**active regions**) on, or immediately below, the solar surface.

supercluster [105]**:** a cluster of **galaxy clusters**; a loose aggregation of thousands of galaxies spread over a region of space up to several hundred million light-years across.

super-Earth [126]**:** an **exoplanet** with a radius of between 1.25 and 2 times that of the Earth.

supergiant [72]**:** an exceptionally large, and extremely luminous, star; supergiants are distributed across the top of the **Hertzsprung-Russell diagram**.

superior conjunction [23]**:** the configuration at which a planet is on the diametrically opposite side of the Sun from the Earth; its **elongation** is then 0°.

superior planet [22]**:** a planet that is further from the Sun than is the Earth.

supermassive black hole [98]**:** a **black hole** with a mass of between a million and several billion solar masses, that may be found in the **nucleus** of a conventional **galaxy** or in an **active galactic nucleus**.

supernova (plural, supernovae) [78, 90]**:** a catastrophic event in which a star blows itself apart and its observed luminosity increases briefly and abruptly by a factor of a million or so. **Type Ia supernovae** reach peak brilliancies equivalent to the luminosity of about 10 billion Suns, about ten times greater than the peak brilliancy of **Type II supernovae**.

supernova remnant [90]**:** the expanding cloud of debris expelled by a supernova.

synodic period [25]**:** the time interval between two successive similar configurations of a planet (e.g., between two successive **oppositions** or two successive **inferior conjunctions**).

T

tail (cometary) [60]**:** dust and ionized gas swept out of the **coma** (head) of a **comet** as it approaches, and begins to recede from, **perihelion**.

temperature fluctuations [116]**:** marginally warmer and cooler patches in the **cosmic microwave background radiation** that indicate the presence of marginally denser regions from which came the seeds from which galaxies and larger-scale structures eventually formed.

terrestrial planet [46]**:** a planet composed primarily of rocks and metals that is broadly similar in nature to the Earth; in the **Solar System**, the planets Mercury, Venus, Earth and Mars.

thermosphere [48]**:** a tenuous layer in the outer atmosphere, which is heated to high temperatures by short-wave (ultraviolet and x-ray) radiation from the Sun (see also, **ionosphere**).

total eclipse [27]**:** the phenomenon that occurs when the Moon passes directly in front of the Sun and completely hides its visible disc; in the case of the Moon, a total eclipse occurs when the whole of the Moon lies within the cone of dark shadow cast by the Earth (see **umbra**).

transit [24, 125]**:** the passage of a smaller body in front of a larger one (e.g. when an **inferior planet** crosses the face of the Sun or an **exoplanet** passes in front of its host star, blocking out a small proportion of its light).

transverse velocity [73]**:** the component of a body's velocity that is directed at right angles to the observer's line of sight (see also, **radial velocity**).

triple-alpha reaction [87]**:** a two-stage nuclear fusion reaction in which three helium nuclei ('alpha particles') combine to form a helium nucleus.

tropical year [19]**:** the time interval between two successive occasions on which the Sun arrives at the **vernal equinox**; the period of the annual recurrence of the seasons. Its duration is 365.2422 days.

troposphere [47]**:** the lowest layer of the Earth's atmosphere, within which temperature decreases with increasing altitude, and most of the clouds and weather systems occur.

Type Ia supernova [90]**:** a type of **supernova** that is believed to occur when nuclear **fusion** reactions are triggered abruptly in the interior of a **white dwarf**, resulting in its complete destruction; about ten times brighter than a **Type II supernova**.

Type II supernova [90]**:** a **supernova** that occurs when the core of a massive star collapses to form a neutron star, and the rest of the star's material (the 'envelope') is blasted out into space; also known as a 'core collapse' supernova.

U

ultraviolet (radiation) [134]**:** **electromagnetic radiation** with **wavelengths** shorter than visible light, which lie within the range 10–390 **nanometres** (nm).

umbra (of a shadow) [27]**:** the central cone of complete darkness in the shadow cast by an opaque body; at any point within the umbra, the illuminating body (e.g., the Sun) will be completely hidden.

umbra (of a sunspot) [36]**:** the darker, cooler, central region of a **sunspot**.

Universal Time (UT) [17]**:** a system of time measurement equivalent to the mean solar time as measured by an observer located on the Greenwich **meridian** (the meridian that passes through the old Royal Observatory in Greenwich, England); formerly, and commonly, known as Greenwich Mean Time (GMT).

universe [1]**:** the totality of everything that exists, including matter (in all its forms, from fundamental particles to galaxies and larger-scale structures), radiation, energy and space itself (see also, **observable universe**).

upper transit [15]**:** *see* **culmination**.

V

vacuum energy [122]**:** a form of background energy in space that permeates the whole **universe** (*see* also, **cosmological constant**).

variable star [76]**:** a star that varies in brightness periodically, irregularly or abruptly.

vernal equinox [13]**:** the point of intersection between the **ecliptic** and the **celestial equator** at which the Sun passes from south to north of the celestial equator. The Sun arrives at this point on or around 21 March each year.

very long baseline interferometry (VLBI) [143]**:** a technique in which very widely separated individual radio dishes (sometimes hundreds or thousands of kilometres apart) are used to make up an **interferometer**.

Virgo cluster [103]**:** the nearest major **galaxy cluster**; located at a distance of about 50 million light-years, in the direction of the constellation of Virgo, it contains about two thousand galaxies.

W–Z

wavelength [134]**:** the distance between two successive crests in a wave motion (e.g. a light wave).

white dwarf [68, 72, 89]**:** a star of low luminosity, but relatively high surface temperature, that has run out of nuclear 'fuel' and has contracted to a radius comparable to that of the Earth.

WIMP [117]**:** an acronym for Weakly Interacting Massive Particle; a hypothetical type of elementary particle that does not interact with electromagnetic radiation or via the strong nuclear force that acts on baryons. WIMPs are the front-running candidates to account for **dark matter**, but, as yet, none have been detected.

winter solstice [15]**:** the point on the **ecliptic** at which the Sun is at its greatest southerly **declination**; the Sun reaches this point on or around 22 December each year.

zenith [11]**:** the point on the celestial sphere that is directly overhead.

Further reading

The Cosmos: Astronomy in the New Millennium (2013) by Jay M. Pasachoff and Alex Filippenko, Cambridge University Press.

Introducing the Planets and their Moons (2014) by Peter Cattermole, Dunedin Academic Press.

Near-Earth Objects: Finding Them Before They Find Us (2012) by Donald K. Yeomans, Princeton University Press.

Exoplanets – Finding, Exploring and Understanding Alien Worlds (2012) by Chris Kitchin, Springer.

Stars and their Spectra: An Introduction to the Spectral Sequence (2011) by James B. Kaler, Cambridge University Press.

Life and Death of Stars (2013) by Kenneth R. Lang, Cambridge University Press.

Revealing the Heart of the Galaxy – The Milky Way and Its Black Hole (2013) by Robert H. Sanders, Cambridge University Press

An Introduction to Galaxies and Cosmology (2004) by Mark H. Jones and Robert J. Lambourne, Cambridge University Press.

Dark Side of the Universe (2007) by Iain Nicolson, Canopus Publishing Limited/ Johns Hopkins University Press.

Planetary Geology 2nd Edition (2014), by Claudio Vita-Finzi and Dominic Fortes, Dunedin Academic Press.

Patrick Moore's Data Book of Astronomy (2014) by Patrick Moore and Robin Rees, Cambridge University Press.

Useful Websites

www.astrosociety.org: Astronomical Society of the Pacific

aas.org: American Astronomical Society (professional organization)

www.ras.org.uk: Royal Astronomical Society

www.britastro.org: British Astronomical Association

www.popastro.com: Society for Popular Astronomy

www.eso.org: European Southern Observatory

www.noao.edu: National Optical Astronomy Observatory

www.nasa.gov: NASA (general site for missions, news, results)

hubblesite.org: Hubble Space Telescope

www.esa.int: European Space Agency

www.spacetelescope.org: ESA–Hubble Space Telescope site

www.eso.org: European Southern Observatory

planetquest.jpl.nasa.gov: Planet Quest (search for exoplanets site)